Lecture Notes of the Unione Matematica Italiana

Volume 30

Yiheng Dong • Marco Martens • Liviana Palmisano

One-Dimensional Dynamics

From Poincaré to renormalization

Springer

Unione Matematica Italiana

Yiheng Dong
Mathematics Department
Stony Brook University
Stony Brook, NY, USA

Marco Martens
Mathematics Department and Institute for
Mathematical Sciences
Stony Brook University
Stony Brook, NY, USA

Liviana Palmisano
Department of Mathematics
KTH Royal Institute of Technology
Stockholm, Sweden

ISSN 1862-9113 ISSN 1862-9121 (electronic)
Lecture Notes of the Unione Matematica Italiana
ISBN 978-3-031-94881-7 ISBN 978-3-031-94882-4 (eBook)
https://doi.org/10.1007/978-3-031-94882-4

This Springer imprint is published by the registered company Springer Nature Switzerland AG
The registered company address is: Gewerbestrasse 11, 6330 Cham, Switzerland

To

周岩*(Zhou Yan) and* 董立国*(Dong Liguo)*
Otto and Els
Giuseppe and Lucrezia

A Lucrezia, perché io sono quello che tu eri e saresti stata.

Preface

This book has a long history. The first draft was written in the summer of 2016 as lecture notes for a one month intense introductory course into dynamics at the Institut de Mathématiques et de Sciences Physiques (IMSP), Benin. The aim of the book is to give an introduction to the main concepts in modern dynamics. The Herman Theorem on rigidity of circle diffeomoprhisms is used to achieve this goal. Indeed we reprove this sophisticated theorem with modern techniques involving renormalization and in the meantime discuss most of the important concepts used in dynamics.

The one-dimensional context of the Herman Theorem reduces the technical difficulties and allows to present tools and theories to students with modest background. The book can definitely be used on master level.

The final version of the book has been possible thanks to colleagues and students who read and commented on the manuscript throughout the years. We would especially like to thank Selim Ghazouani, José São João, Luca Marchese, Corinna Ulcigrai, and Heather Werth.

Stony Brook, NY, USA — Yiheng Dong
Stony Brook, NY, USA — Marco Martens
Stockholm, Sweden — Liviana Palmisano

Preface

Contents

Chapter 1
Introduction

1.1 History

The motion of planets has been a source of enduring fascination for people throughout history. Incremental progress has been made in our understanding of the solar system over time. The earliest breakthroughs predominantly came from meticulous observations and were driven by the practical need to predict events essential for everyday life, such as those related to agriculture and farming. Initially, a comprehensive model was absent. It was only during the Greek period that the first formal attempt at explaining planetary motion emerged, proposing that planets follow circular orbits centred around the Earth.

Subsequent advancements didn't occur until the first decades of the sixteenth century when N. Copernicus provided a geometric representation of planetary motion, asserting that planets follow circular orbits around the sun, [Cop52]. This representation is called the heliocentric model of the solar system. This geometric model, together with mathematical tools, enabled him to make accurate predictions for the celestial movements across the sky. Indeed, if one observes a given planet, say Mars, each day at the same time, say midnight, one sees the planet changing position from night to night. Surprisingly, this movement is not, in the case of Mars, monotone from west to east. There are periods when Mars moves in the opposite direction, from east to west. This is called the retrograde motion of Mars (Fig. 1.1). The geometric model explained these complex paths taken by planets like Mars.

Copernicus fundamentally reshaped our understanding by asserting that planets trace circular paths around the sun, dispelling the Earth as the centre of the universe. Naturally, this raised a question: Is this assertion accurate, and if so, why?

A preliminary experimental response came in the beginning of the seventeenth century from G. Galilei, who turned his telescope toward the sky and saw the elliptical orbits of Jupiter's moons. Galileo's observation was ground breaking, serving as evidence that not everything turns around the Earth, [Gal89].

Y. Dong et al., *One-Dimensional Dynamics*, Lecture Notes of the Unione Matematica Italiana 30, https://doi.org/10.1007/978-3-031-94882-4_1

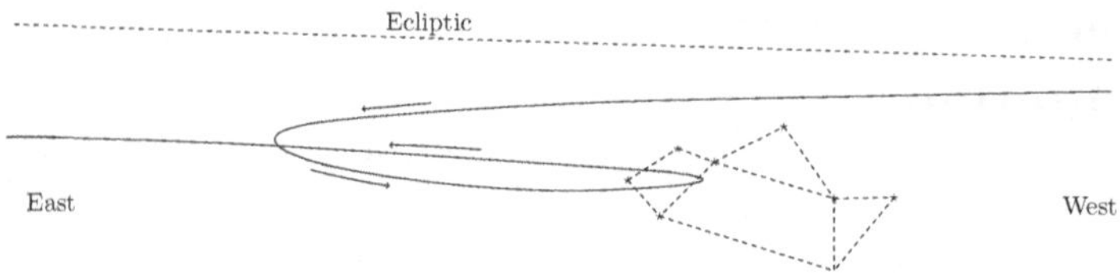

Fig. 1.1 Retrograde motion of Mars

The study by Copernicus and Galileo of the motion of the planets was primarily experimental and descriptive. A century later, I. Newton presented an explanation for the elliptical orbits of the planets rooted in two fundamental elements, one of physical and one of mathematical nature. The cornerstone of his explanation was the concept of gravitational force, which dictates that two masses attract each other according to a precise law. This gravitational force operates among all the components within the solar system.

To predict the effects of gravitational forces on planetary motion, Newton invented a mathematical tool known as differential equations. The acceleration $a = d^2x/dt^2$ of a given planet with position $x(t)$ at moment t, is given by the equation

$$F = m \cdot a,$$

where m is the mass of the planet and F is the gravitational force acting on the planet. Given an initial position and velocity of a planet, solving Newton's differential equation provided the exact position of the planet at any given time. Specifically, he applied this method to the simplified model of a single planet orbiting the sun. The resulting solution from the differential equation was an elliptical orbit, thereby supporting the Copernican geometric model, [New99].

One of the most remarkable achievements of Newton's approach was its inherent determinism. It enabled the prediction of future events. Namely, starting from an initial state of the system, solving the differential equations makes it possible to deduce the future from the present circumstances.

The study of Newton's differential equations became a focal point of extensive research. Unfortunately, solving differential equations is often a difficult task, frequently practically impossible. The solutions available at the time tended to exhibit cyclic or periodic behaviour. This led to a widespread belief that all systems followed periodic patterns, as seen in the case of a single planet orbiting the sun, pendulum motion, or even the changing seasons.

In the latter part of the nineteenth century, a question emerged regarding the enduring stability of the solar system. Was it possible that the solar system's behaviour would remain constant indefinitely? Could we rule out the possibility of significant alterations in planetary orbits over an extended period, or even the ejection of a planet from the system? These questions were posed within the context of a prestigious mathematics prize competition commemorating King Oscar II of

Sweden's 60th birthday and were deliberated by some of the era's most influential mathematicians. The exact question was articulated as follows:

"For a system of arbitrarily many mass points that attract each other according to Newton's law, assuming that no two points ever collide, find a series expansion of the coordinates of each point in known functions of time converging uniformly for any period of time."

H. Poincaré took up the challenge, ultimately winning the prize, although he did not arrive at a definitive answer. To this day, the question of the solar system's stability remains unresolved. However, Poincaré made a significant contribution by constructing a mechanical system with non-cyclic orbits that exhibited chaotic behaviour, [BG94, Poi90, Poi92]. A noteworthy aspect of Poincaré's approach to the problem was his deliberate avoidance of solving the equations. Instead, he focused on studying the qualitative characteristics of the solutions without engaging in explicit numerical computation.

1.2 Dynamics

Dynamics studies the evolution of systems over time. Often dynamical systems are given by differential equations. Instead of studying whole solutions of the differential equations, i.e. curves, Poincaré showed that the essential aspects of the system can often be described by a map on a state space X

$$f : X \to X.$$

An example of this method is discussed in Sect. 1.3.

Consider the dynamical system given by the map f. A point $x \in X$ represents the state of the system. For example, the point x could include the position of the planets and their velocity at a given moment. Indeed, X can be a high dimensional manifold. The map f tells how the state changes after one second, the new state will be $f(x)$. By iterating the map one finds the whole future evolution of the system starting at the initial state x. In essence, iterating the map f is similar to finding the solution of a differential equation starting at an initial point. The only difference in our situation is that we consider discrete time, while the time used in differential equations is continuous. The idea is the same, i.e. the starting point determines the whole future.

The future of a starting point is called the *orbit* of x, namely

$$O(x) = \{x, f(x), f^2(x), \cdots, f^n(x), \cdots\},$$

where

$$f^n = f \circ f \circ \cdots \circ f,$$

is the nth iterate of f.

In many applications one wants to predict what happens tomorrow or even in a couple of seconds. Short term predictions are often very important. In our study of dynamics we will only concentrate on the long term behaviour of orbits. Namely we study the tail of the orbit, i.e. the points $f^n(x)$ with n very large. The initial parts of the orbit is not of our interest. The idea, the hope, is that after a long time the orbits will settle down in an equilibrium and this equilibrium is independent of the starting point.

The simplest situation is when the equilibrium is a *fixed point* $p \in X$, i.e.

$$f(p) = p,$$

or a *periodic point*, namely a point whose orbit is finite

$$O(p) = \{p, f(p), f^2(p), \cdots, f^n(p) = p\}.$$

However, in general, the equilibria one can observe are more complicated *invariant* sets $A \subset X$. Invariant means $f(A) = A$. They can be manifolds, Cantor sets or sets with even more intricate structure.

Instead of equilibrium we will use the terminology *attractor*. An attractor of the system is an invariant set $A \subset X$ such that most orbits accumulates at this set, namely, for x in a positive measure set,

$$A = \bigcap_{n \geq 1} \overline{O(f^n(x))}. \tag{1.2.1}$$

The collection of all points with property (1.2.1) is called the *basin of attraction*. Observe that the orbit of a point in the basin of attraction accumulates at the attractor. In particular, the tails of these orbits are indistinguishable from the attractor. Hence, the attractor contains all information on the long term behaviour of orbits. The attractor contains the essence of the system.

We explain now why we chose to concentrate on the long term behaviour of the system. If one observes a real world system, say the solar system, it has been evolving already for a long time previously. Starting to study it today really means that we are looking at the tails of orbits of the system which are indistinguishable from the attractor.

There are two central questions in dynamics. The first question involves the state space and can be formulated as follows.

State Space Question: Does the system have an attractor and what are its topological, geometrical, and measure theoretical properties, and maybe even, what are its algebraic properties?

In order to state the second question one needs to introduce the concept of parameters. Often, systems have parameters. Consider for example, a double pendulum. These pendulums consist of two rods, one hinged at a fixed point, the other hinged at the end of the first rod, see Fig. 1.2. The motion of such a double pendulum can be rather complicated. It is almost funny to watch, especially when

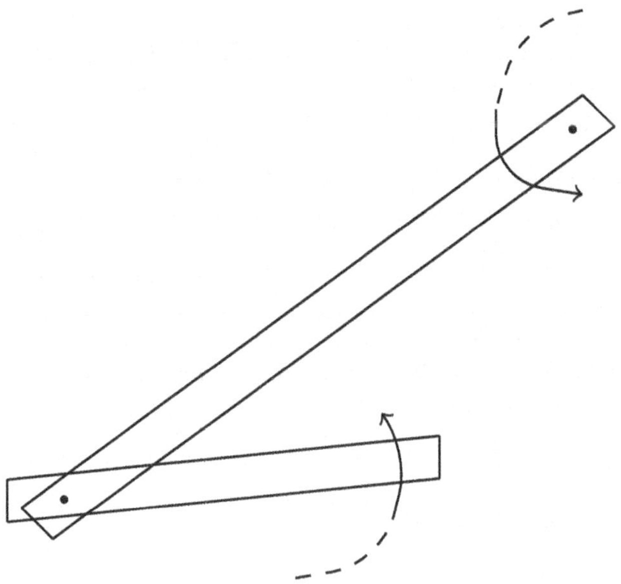

Fig. 1.2 A double pendulum

one is used to the solemn movement of a pendulum, [Tro07]. This spectacle can be described by a differential equation. The length of each rod will appear in this differential equation as fixed parameters, to be adjusted for the given double pendulum. Another example could be the strength of an external force acting on the system, e.g. the energy of the sun shining on earth effecting the weather. In this case, the energy of the sun represents another parameter.

These parameters are often not exactly known. Moreover, it occurs sometimes that by changing the parameters, even a little bit, the behaviour of the system changes dramatically. This means that if one finds an attractor and one wants to determine its relevance, one also needs to understand for which values of the parameters the corresponding attractors are the *same*.

What does it mean that two attractors are the same? Consider two dynamical systems $f, g : X \to X$ with attractors A_f and A_g. The attractors are the same if there exits a coordinate change, i.e. a homeorphism $h : A_f \to A_g$, such that h identifies the orbits in the attractor of f with the orbits in the attractor of g, namely

$$h \circ f = g \circ h.$$

The ideal situation is when the attractors are the same within an open set of parameters. In this case we say that the attractor is *structurally stable*. Many relevant attractors are not structurally stable but sometimes the set of parameters in which the attractors are the same form a submanifold. The set of parameters for which the attractors are the same is called the *class of the attractor*.

One expects that attractors which play a role in nature, should persist along rather large classes. Otherwise they would not be observable. The second central question in dynamics can be stated as follows.

Parameter Space Question: The class of a structurally stable attractor is by definition an open domain in parameter space. What is the structure of the class when the attractor is not structurally stable? What are the topological, geometrical, and measure theoretical properties, of the classes of attractors? Where does the attractor, with its properties, survive?

There is definitely no general theory to address these questions for all systems. It is more as if we are in the very beginning of the development of such a theory. However, the little we do understand shows that dynamical systems can exhibit fascinating properties. What we do understand falls into two groups, hyperbolic and non-hyperbolic systems.

1.2.1 Hyperbolic Systems

The simplest example of a hyperbolic system is an attracting periodic point, a sink. In 1960, S. Smale, walking on the beach of Copacabana in Rio, discovered another example of a hyperbolic system, more complicated than a sink. He found what then became the prototypical hyperbolic system: the Smale Horseshoe, see Fig. 1.3. Smale constructed the Horseshoe as an transparent example of hyperbolic systems containing infinitely many periodic orbits, [Sma65, Sma80]. The topological structure of the Horseshoe is the product of two Cantor sets

$$X = C \times C,$$

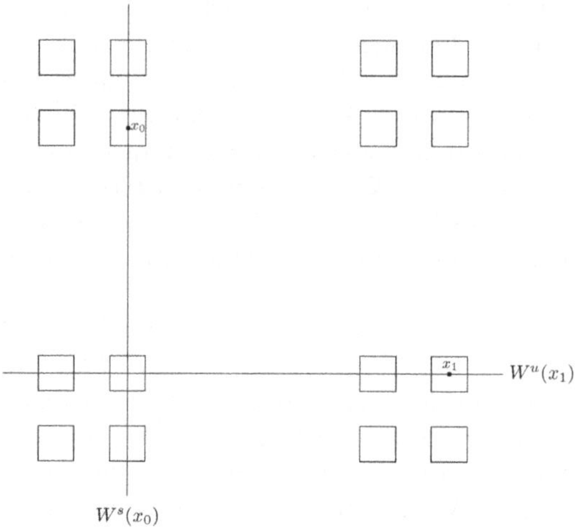

Fig. 1.3 The horseshoe

where C is a Cantor set in the real line. One should think about the vertical Cantor sets for $x = (c_1, c_2) \in X$,

$$W^s(x) = \{c_1\} \times C$$

as the stable sets. Namely, the orbit of any $y \in W^s(x)$ behaves the same as the orbit of x, i.e.

$$|f^n(x) - f^n(y)| \to 0$$

exponentially. To the contrary, the horizontal Cantor sets for $x = (c_1, c_2)$,

$$W^u(x) = C \times \{c_2\}$$

are the unstable sets. Namely, the orbit of any $y \in W^u(x)$ diverges exponentially fast from the orbit of x. An essential property of hyperbolic systems is the presence of unstable and stable behaviour and a (local) product structure as in the Smale Horseshoe.

Sometimes people use the word *chaos* to say that the behaviour of the system is sensitive to initial condition. This, in turn, means that one can change the starting point only slightly and the orbit of the new point will behave very differently from the orbit of the original starting point. One can see that the Smale Horseshoe is chaotic in this sense.

The State Space and Parameter Space questions have answers in the context of hyperbolic dynamics. Namely, the topological properties of hyperbolic systems are completely understood. They can be described with so-called symbolic dynamics. Also the measure theoretical aspects are well understood. The hyperbolic systems are structurally stable, i.e. systems which are close enough to a given hyperbolic one are also hyperbolic and, up to a coordinate change, are the same as the given one. The class of a hyperbolic attractor is an open set in parameter space.

Although the topological properties of a hyperbolic attractor do not change throughout its class, the geometrical properties do change. In particular, the coordinate change between two hyperbolic attractors might just be a homeomorphism and not smooth at all. For example, one can change the Hausdorff dimension of the Horseshoe by slightly changing the parameters. Surprisingly, there are non-hyperbolic attractors whose geometry does not change after perturbation within its class. The coordinate changes are essentially smooth. This phenomenon is a central theme of the book.

There is an extensive literature on hyperbolic systems. One can find introductions to hyperbolicity within general references to dynamics. For example [BS02, GH83, KH95, Mn87, PM82], and [PT93].

1.2.2 Non-hyperbolic Systems

The theory for non-hyperbolic systems is fragmented. There is definitely not a general non-hyperbolic theory. We will discuss some of the results which are relevant as background for the book.

If the state space is one-dimensional, the circle or the interval, there is a mature description of the dynamics. The topology of the attractors is very well understood using again symbolic dynamics, in this case in the form of kneading theory, [MS93]. Typical systems have invariant measures which describe the measure theoretical aspects of the systems, [ALM03, AM03, AM05a, AM05b, AM05c, Lyu00], and [Lyu02]. The classes of systems are finite codimension manifolds, [ALM03, Cla14, CS23].

Although the State Space and Parameter Space questions are adequately answered, there is a surprising phenomenon in non-hyperbolic one dimensional dynamics. In the hyperbolic context, as we saw before, one can deform the attractor and change its geometrical properties. Surprisingly, there are non-hyperbolic attractors, the so-called *infinitely renormalizable* attractors, whose geometry cannot be deformed. The coordinate changes between such attractors is smooth. On small scale one can not deform the geometry. In particular, the topology of these infinitely renormalizable systems determines the small scale geometrical properties of the system. This is a central theme in one-dimensional dynamics, known as *rigidity*. The following list of references is long but not complete, [Avi13, Far99, FM99, FM00, FKS82, GMM18, GM17, Her79, KT07, KY06, Lan87, Lan88, Mel87, MP99, Öst+83, Ran88a, Ran88b, She82, Yam99, Yam01, Yam02, Yam03].

The theory behind rigidity is quite delicate. There are topological aspects which play a role but also the smoothness of the systems is essential. In lower smoothness, say when the systems are only C^1, rigidity does not occur, [Cha+09]. It might also occur that there are finitely many geometrical invariants which determine when two attractors are smoothly conjugated, [LM93, MP22].

The study of rigidity is not at all finished. There are still many unexplored areas, even in one-dimensional dynamics. For example, one expects rigidity in systems similar to the Lorenz Butterflies, [GH83, MW14, MW16].

In the following sections and later in the book we will give a more detailed explanation of rigidity for the simplest type of dynamical systems, i.e. circle diffeomorphisms. Let us briefly discuss the State Space and Parameter Space questions for systems in higher dimension.

If the state space is two-dimensional the theory is only very partially developed. Central examples are the maps in the so-called *Hénon family*, $F_{a,b} : \mathbb{R}^2 \to \mathbb{R}^2$

$$F_{a,b} : (x, y) \mapsto (a - x^2 - by, x),$$

with parameters $(a, b) \in \mathbb{R}^2$. There are examples of Hénon maps exhibiting attracting periodic orbits, for example when $a = 0$ and $|b| < 1$, the corresponding Hénon map has $(0, 0)$ as an attracting fixed point. In the case when a is large and b small, the corresponding Hénon map is hyperbolic and it has a Horseshoe. Moreover, the Hénon family also contains non-hyperbolic maps, which, following our present understanding, one can group in three branches.

There are Hénon maps with infinitely many periodic attractors, [BP23, Dav91, GST08, New74], [Rob83, Str81]. These are examples of the *Newhouse phenomenon*. Newhouse was the first to construct two-dimensional diffeomorphisms with infinitely many coexisting periodic attractors, [New99]. These maps that Newhouse constructed are not hyperbolic. In particular, they are not structurally stable. There are still many questions unanswered concerning the Newhouse phenomenon. For example, related to the Parameter Space question, does the the Newhouse phenomenon occur with positive measure in parameter space? A partial answer is given in [Ber16].

There is a set of positive measure in parameter space where the attractor of the corresponding Hénon map is not hyperbolic. These maps were constructed by Benedicks and Carleson, [BC91]. There are still many questions open related to these attractors. The measure theoretical aspects are understood but the topological structure of these attractors requires further study.

The third branch are the maps with the simplest topological structure. These are the maps at the *boundary of chaos*. A system at the boundary of chaos can be deformed such that the perturbed map has only finitely many periodic attractors but also, a map at the boundary can be deformed such that it has attractors which are much richer than periodic attractors.

The maps at the boundary of chaos have an attractor which is a Cantor set. Indeed, the topological and measure theoretical aspects of these attractors are well understood. Also in parameter space, the class, i.e. the boundary of chaos, is a codimension one manifold. The State Space and Parameter Space questions have answers, [CEK81b, CPT20, DCLM05, GST89, LM11a, LM11b].

The geometry of these Cantor attractors at the boundary of chaos, exhibits a fascinating phenomenon. They are not rigid, the coordinate changes between two such attractors is at most Hölder. This phenomenon was a surprise when it was discovered because their one-dimensional counterparts are rigid. Rigidity in this two-dimensional context is more delicate. Namely, in almost all points of the attractor, the coordinate change is smooth. This phenomenon is called probabilistic rigidity, [LM21].

These different examples of non-hyperbolic dynamics have many aspects. Some are well understood but there are also many related open questions. These examples are only in one and two dimensional state spaces. One should not expect that more general systems, i.e. higher dimensional systems, behave similar as the one and two dimensional systems. Recall that in two dimensional state spaces there is the Newhouse phenomenon and probabilistic rigidity which does not occur in one-dimensional dynamics. One should be awaiting other fascinating phenomena in more general dynamics. Probably, two-dimensional dynamics did not yet reveal all its secrets.

As often happens in mathematics, one needs tools to progress. The goal of this book is to provide some of the tools used to develop our understanding of the examples above. We will discuss in detail some of the phenomena mentioned above in the simplest context, namely that of circle diffeomorphisms. Even in this simplest context, the theory is sophisticated. However, the technical ingredients are relatively elementary. The main motivation of this book is to expose the reader to many tools which might be used, or might be a starting point, in more general context.

1.3 Circle Diffeomorphisms

In this section, we motivate the study of circle diffeomorphisms which describe the (quasi) periodic motion of the planets. This is an example of Poincaré's method to reduce the study of flows to the study of maps.

Let us start by examining the simplest scenario of one planet orbiting the sun in a circular orbit. In this configuration, the planet's orbit lies in a plane, and its position can be described by an angle using polar coordinates. The only variable is the angle θ_1 whose dynamics is entirely encapsulated by the solution to the following differential equation:

$$\frac{d\theta_1}{dt} = \alpha_1$$

where α_1 is the so called *angular velocity* reflecting the fact that the planet is moving with constant speed. The previous equation has solution

$$\theta_1(t) = \theta_1(0) + t\alpha_1 \bmod 2\pi.$$

In this simple situation, the planet moves in a circle with constant speed, and the state of the planet is described by an angle. We say that the state space of the planet is the unit circle $\mathbb{S}^1$ in the plane.

Now, imagine the presence of another planet, entirely independent of the one mentioned earlier, both orbiting the sun in circular paths. We can assume that these two planets are exceptionally small so that we can omit the force in between them, and the sole forces at play in this system are those acting between each planet and the sun. The dynamics of the second planet is then described by an angle θ_2 and equation

$$\frac{d\theta_2}{dt} = \alpha_2$$

having solution

$$\theta_2(t) = \theta_2(0) + t\alpha_2 \bmod 2\pi.$$

As before, the state space of the second planet is the unit circle $\mathbb{S}^1$ in the plane.

Fig. 1.4 A linear torus flow

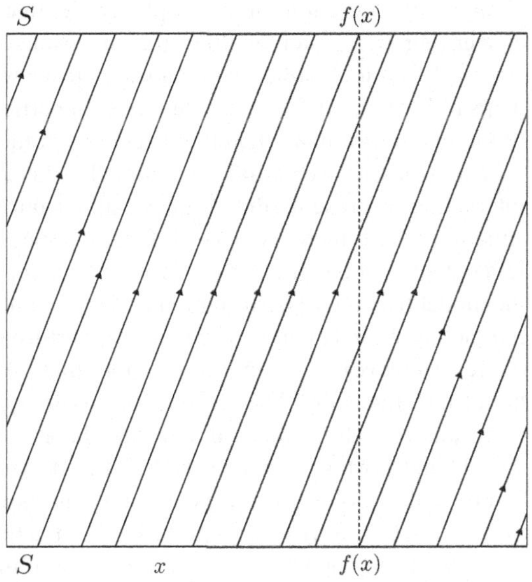

The state of two planets circulating the sun is given by a pair of angles. The state space of the entire system is the product of two circles, namely a torus,

$$\mathbb{T}^2 = \mathbb{S}^1 \times \mathbb{S}^1.$$

Moreover, the state of our system evolves according to the differential equation

$$\frac{d\theta}{dt} = \alpha \tag{1.3.1}$$

where $\theta = (\theta_1, \theta_2)$ and $\alpha = (\alpha_1, \alpha_2)$. Equation (1.3.1) represents a linear flow on $\mathbb{T}^2$, see Fig. 1.4.

We consider now the circle $S = \{\theta_2 = 0\}$. Observe that all orbits of the flow cross this circle transversally. Moreover all orbits starting from S come back to S. A loop S with these two properties is called a *section*. The orbits of the flow can be fully understood by the way they repeatedly return to the section. The *first return map* to S, $f : S \to S$ is then defined in a point $x \in S$ as the first point in S where the orbit of x comes back to S, see Fig. 1.4. The return map f, in this case, is a rotation over angle $2\pi\alpha_1/\alpha_2$, namely

$$f(x) \to x + 2\pi \frac{\alpha_1}{\alpha_2} \mod 2\pi.$$

The first return map is also called the Poincaré map.

We began with a system described by differential equations and reduced our analysis of the solutions of the differential equation to the study of the orbits obtained by iteration of a simple circle map, a rotation. Now, all the details concerning the system's flow are encapsulated within the dynamic of f. This approach is a technique that enables us to view dynamical systems as iterations of maps rather than differential equations. In particular, if one understands the attractor of the first return map, then the behaviour of the original flow can be reconstructed.

The scenario we initially examined, where two planets orbit the sun without interacting, represents the simplest, albeit unrealistic, case. Now, let's consider a situation where there exist weak forces among the planets, which are significantly less potent than the gravitational forces between the planets and the sun. In this case, the model becomes a perturbation of the previous one. Often a solar system with two interacting planets can be viewed as a perturbation of the linear motion described earlier. However, it's important to note that while this is often the case, it doesn't hold true universally. The reasons for this are quite intricate and depend on the ratio of α_1 and α_2. This is known as KAM theory, as introduced by A. Kolmogorov, V. Arnold and J. Moser, [Arn63, Kol54, Lla01, Mos62, Pös01].

Given the book's objectives, we make the assumption that a solar system with two planets can be described as a perturbation of a linear flow on a torus. In this scenario, the first return map becomes a perturbation of the rotation, a circle diffeomorphism.

The upcoming chapters will be exclusively dedicated to formally defining and exploring circle diffeomorphisms. These maps are, in fact, the most elementary dynamical systems. Nevertheless, their theory is so intricate that it serves as a platform for illustrating many advanced tools applicable to more general dynamical systems. The primary aim of this book is to equip the reader with a toolbox of instruments that can be effectively employed in much broader contexts, extending well beyond the realm of circle diffeomorphism dynamics. The central tool in this endeavour is renormalization.

Throughout the course, the reader will gain a comprehensive understanding of the theory of circle diffeomorphisms, spanning from Poincaré's early findings to renormalization.

1.4 Outline of the Book

There is an extensive theory for the dynamics of circle homeomorphisms. The book will present its main parts.

Chapter 2: Topology of Circle Homeomorphisms The second chapter discusses the topological properties of circle homeomorphisms. It will be shown that the orbits of a rotation with irrational rotation number α_1/α_2 are dense in the circle. When the rotation number is rational, all orbits are periodic. Moreover, the orbits of rotations with different rotation number are topologically/combinatorially different. In the 1880s, H. Poincaré introduced the *rotation number* $\rho(f)$ for a general circle homeomorphism f, [Poi93]. He showed that for a circle homeomorphism f that does not have periodic orbits, the combinatorial structure of the orbits is completely determined by its rotation number $\rho(f)$. More precisely, there exists a coordinate change h, i.e. a monotone increasing surjective continuous function such that

$$h \circ f = r_{\rho(f)} \circ h,$$

where $r_{\rho(f)}$ is the rotation over angle $2\pi\rho(f)$. The existence of such h, which is called a *semi-conjugacy*, implies that the order of the points in an orbit of f is exactly the same as the order of the points in the orbit of the corresponding rotation. In other words, the combinatorial properties of the orbits of f are the same as those of the orbits of the rotation. The rotation number is the main topological invariant in the theory of circle homeomorphisms. Chapter 2 will introduce the rotation number and discusses its properties.

The rotation number describes the topological properties of orbits in state space. To study parameter space one considers families of circle homeomorphisms and as first question one can ask if all rotation numbers are realized in the family. If this is the case, we say that the family is a *full family*. Chapter 2 addresses which one-parameter families of circle homeomorphisms are full families.

The chapter presents two proofs for the full-family theorem. The classical proof is very specific in the sense that it uses the intermediate value theorem which is not available when the topological invariants are higher dimensional objects and not one-dimensional as in the case of the rotation number. The second proof uses the Thurston method, which is more general than the previous one and can be applied to many different settings far beyond the context of circle homeomorphisms. See for example [MP19, MM01, MS93].

Chapter 3: Topology of Circle Diffeomorphisms The third chapter discusses in more detail the semi-conjugacies between circle homeomorphisms and their corresponding rotation. In 1932 A. Denjoy realized that if $f \in C^{1+bv}$, i.e., f is a C^1 circle diffeomorphism with irrational rotation number and its derivative f' has bounded variation, then the semi-conjugacy h from f to the rotation $r_{\rho(f)}$ is a homeomorphism. In this case the dynamics of f and of the corresponding rotation are topologically equivalent.

Observe that to understand the topology of the systems, Denjoy's Theorem requires that the system is a circle diffeomorphism in C^{1+bv}. These smoothness requirements do play an essential role in one-dimensional dynamics. Indeed Denjoy also constructed examples of C^1 circle diffeomorphisms where the semi-conjugacies are not homeomorphisms, [Den32]. These Denjoy Counterexamples have *wandering intervals*. Other Denjoy Counterexamples, showing the delicacy of the existence of wandering intervals are the C^∞ circle homeomorphisms constructed by G.R. Hall, [Hal81]. These examples have one point where the derivative is zero, they are not diffeomorphisms. The study of the (non-)existence of wandering intervals has a long history. The following list of references is long but not complete [BL89, Guc79, Guc87, Lyu89, MMS92, Mar+90, Mel87, MS88, MS89, Pal15, Pal19, Yoc84b].

Chapter 4: Renormalization A natural question that follows the topological classification of circle diffeomorphisms is the *rigidity* question, which asks about the regularity of the conjugacy. Smoothness of the conjugacy implies that on small scale the geometry of f is the same as the small scale geometry of $r_{\rho(f)}$. In the case of diffeomorphisms with irrational rotation number, the orbits are dense. It would be rather surprising if indeed the combinatorial properties of the orbits, encoded in the rotation number, determine the small scale geometrical properties of the orbits. Early results obtained by V.I. Arnold [Arn61, Arn83] and J. Moser [Mos66] show that indeed for diffeomorphism which are small perturbations of the rotations, and have appropriate rotation number, the conjugacies are differentiable.

M. Herman, [Her79], in 1979, followed by J.C. Yoccoz, [Yoc84a], in 1984 with substantial refinements, proved the celebrated global rigidity theorem stating that if f is an orientation preserving C^3 circle diffeomorphism with appropriate rotation number, then f is C^1 conjugated to a rotation. In particular, a Diophantine rotation number assures the conjugacy to be smooth.

Observe, that higher regularity of the circle diffeomorphism cannot guarantee smooth conjugacy to rotations, and in particular, the assumption on rotation numbers is essential. In fact, Arnold, [Arn61], gave an example of a real analytic circle diffeomorphism with irrational rotation number that is not C^1 conjugated to the rotation. The Arnold examples will be discussed in a later chapter, see Chap. 9.

Nowadays there are different methods available to prove the rigidity of circle diffeomorphisms, see for example [KO89, KS87], and [Sta88]. These methods lead to various refinements and improvements of the original results by Herman and Yoccoz. The method used by J. Stark is *renormalization*. This is the method we will employ here. The renormalization technique can be used to prove rigidity. It has many more applications in the theory of circle diffeomorphisms and it is a technique which has been used far beyond the context of circle diffeomorphisms. The introduction of renormalization is the main motivation of the book. In the case of circle diffeomorphisms one can study renormalization with relatively modest previous background.

What is renormalization? Given a space of systems, for example the space of circle diffeomorphisms, the renormalization operator acts on this space as a microscope. In particular, the renormalization of a circle map, is another circle map which describes the original map on a smaller scale. It replaces the original map by the first return map to a smaller domain. One can iterate the renormalization operator which means that one looks at the original map on smaller and smaller scales. The asymptotic geometrical properties of the attractor of the original map depends on the limit behaviour of the consecutive renormalizations. The attractor of renormalization plays a central role. The idea and hope is that the attractor of renormalization is hyperbolic. For example, in the simplest situations the attractor is a fixed point. The converge of the consecutive renormalizations towards this fixed point plays a crucial role for questions related to rigidity. The unstable set of the fixed point will have consequences in parameter space. An advantage of the renormalization operator is that it allows to study the dynamics in state space and parameter space simultaneously.

Renormalization was first introduced into dynamics by Coullet-Tresser, [TC78], and Feigenbaum, [Fei78, Fei79], in the late 1970s to understand the geometric universality phenomena of unimodal maps. Their main conjecture was that in the context of unimodal interval maps at the boundary of chaos there is a unique renormalization fixed point which is hyperbolic. This conjecture has been proved during the last decades. The main results in unimodal renormalization theory were obtained by Sullivan, McMullen, Lyubich, among others. For relevant references, see e.g. [AL11, BMT03, Eps86], [FMP06, Lan82, Lyu89, Mar98, McM94, McM98, Sul92, VSK84]. There is also an extensive renormalization theory for circle maps, see e.g. [Far99, FM99, FM00, GMM18, GM17, Her79, KT07, KY06, Lan87, Lan88, MP99, Yam99, Yam01, Yam02, Yam03].

Although there is mature one-dimensional renormalization theory, there might be still surprises in one-dimensional dynamics. For example, the theory for Lorenz maps is still in its beginning, [MW14, MW16]. In two dimensional dynamics there are substantial applications of renormalization, see [CEK81a, DCLM05, EKW82, EKW84, GJM16, GJ09, GJ16, HLM12, LM11a, LM11b], [LM21]. Also in this case, we are very far for a complete theory.

We introduce the renormalization operator for circle diffeomorphisms in Chap. 4. The convergence of consecutive renormalization is a central aspects of renormalization. The first step towards convergence is to show that these consecutive renormalizations stay in a bounded set of maps. This boundedness is known as the *a priori bounds*. A priori bounds are a central theme in renormalization theory. For circle diffeomprophisms they are obtained in Chap. 4.

Chapter 5: The Three Ingredients for Convergence of Renormalization The attractor of renormalization of circle diffeomorphims is the curve of rotations. The convergence towards the attractor requires some work. The three ingredients responsible for the convergence are introduced in Chap. 5. Namely, convergence from diffeomorphisms to Möbius maps, the convergence from Möbius maps to affine maps, and finally, the convergence of affine maps to rotations.

Chapter 6: The Collaboration of the Three Ingredients These three ingredients collaborate and cause the convergence of renormalizations. This is explained in Chap. 6. One can use this strategy to show convergence as a guideline for generalization to more general contexts. See for example [GU21, Mar98].

Chapter 7: Exponential Convergence The consecutive renormalizations might only converge to the affine maps. However, a properly chosen subsequence of renormalizations, the *fast renormalizations*, converge exponentially to the rotations. This is explained in Chap. 7. The examples of maps with poor converge properties are constructed in Chap. 9. In the case when the irrational rotation number is of bounded type, i.e., the entries in the continued fraction expansion are bounded by a constant, then the sequence of all renormalizations does converge exponentially towards the rotations. This is also discussed in Chap. 7.

Chapter 8: Geometry of Circle Diffeomorphisms The rigidity of circle diffeomorphisms is the subject of Chap. 8. To avoid poor converge of the sequence of renormalizations, as in the examples of Chap. 9, one needs to impose a condition on the rotation number. In particular, it is a condition in terms of its continued fraction expansion. Under this condition the geometrical properties of the whole orbit are described asymptotically by the limits of renormalization, i.e. by the corresponding rotations. The rigidity Theorem follows.

Chapter 9: Non-rigidity This chapter presents example of maps with poor convergence properties of their renormalizations. The first example is Arnold's non-rigidity example. There exist analytic circle diffeomorphisms with irrational rotation number for which there is a subsequence of renormalizations which does not converge to the rotations. Indeed, this example is also not smoothly conjugated to the corresponding rotation.

The second example is an affine circle map whose rotation number assures rigidity for circle diffeomorphism, but there is a subsequence of renormalizations of this map which does not converge exponentially to the rotations.

Chapter 10: The Renormalization Horseshoe The renormalization operator is used to describe certain types of non-hyperbolic dynamics. The success of renormalization relies on its hyperbolicity. As an operator on the space of systems it is itself a dynamical system which, in the cases we do understand, has a hyperbolic nature. Observe, circle diffeomorphisms are non-hyperbolic, while their renormalization operator it is hyperbolic. This is the key behind the applications of renormalization.

One has to be careful, the definition of hyperbolicity requires the system to be smooth, see [BS02, Mn87, PM82]. Usually, in a smooth context, renormalization is not differentiable. In particular, Chap. 10 shows that indeed renormalization of smooth circle maps is not differentiable.

However, although renormalization of circle diffeomorphisms is not differentiable, fast renormalization behaves as if it were hyperbolic. Namely, there is a stable lamination whose leafs are the topological classes and the attractor consists of one unstable leaf. The *topological hyperbolicity* of circle renormalization is discussed in Chap. 10. The hyperbolicity of renormalization is the main pillar of the theory.

1.5 Smoothness

In dynamics one has to be careful about the meaning of the word *smooth*. The main subject in this book are circle diffeomorphims. It is remarkable that the behaviour of the diffeomorphims depends on their smoothness. For example, if the map is only $\mathcal{C}^1$ there might be wandering intervals. These wandering intervals do not exit when the map is $\mathcal{C}^2$. The rigidity phenomena occur when the diffeomorphisms are $\mathcal{C}^3$. The theory of rigidity for circle diffeomorphims does not require the maps to be analytic or $\mathcal{C}^\infty$. Generally speaking we will use the word smooth as referring to $\mathcal{C}^3$, if not specified otherwise.

The degree of smoothness plays a delicate role in dynamics.

1.6 Exercises

Exercise 1.6.1 Let us consider the middle third Cantor set $C_{\frac{1}{3}} \subset [0, 1]$ and similarly $C_{\frac{1}{4}} \subset [0, 1]$ obtained by removing at each step the middle interval of relative length $\frac{1}{4}$. Having corresponding pieces of different sizes, $C_{\frac{1}{3}}$ and $C_{\frac{1}{4}}$ are not geometrically the same. Are they topologically the same? In other words, does there exists a continuous bijection $h : [0, 1] \to [0, 1]$ such that $h(C_{\frac{1}{3}}) = C_{\frac{1}{4}}$.

Exercise 1.6.2 Is the homeomorphism h constructed in Exercise 1.6.1 differentiable? Is h^{-1} differentiable?

Exercise 1.6.3 Let $C_{\frac{1}{3}}$ be the middle $\frac{1}{3}$ Cantor set and $C_{\frac{1}{3}}^{\frac{1}{n}}$ be the Cantor set where the ratio used to construct the $n + 1$ level is

$$\left(1 + \frac{1}{n}\right)\frac{1}{3}.$$

These Cantor sets are topologically the same. Can you find a homeomorphism $h : [0, 1] \to [0, 1]$ such that $h(C_{\frac{1}{3}}) = C_{\frac{1}{3}}^{\frac{1}{n}}$ and h is differentiable?

Exercise 1.6.4 Let $C_{\frac{1}{3}}$ be the middle $\frac{1}{3}$ Cantor set and $C_{\frac{1}{3}}^{\frac{1}{n^2}}$ be the Cantor set where the ratio used to construct the $n + 1$ level is

$$\left(1 + \frac{1}{n^2}\right)\frac{1}{3}.$$

Can you find a homeomorphism $h : [0, 1] \to [0, 1]$ such that $h(C_{\frac{1}{3}}) = C_{\frac{1}{3}}^{\frac{1}{n^2}}$ and h is differentiable?

Exercise 1.6.5 Represent the torus $\mathbb{T}^2$ by a unit square identifying the horizontal sides and the vertical ones in the proper fashion. On $\mathbb{T}^2$ draw the orbits of the three linear flows

$$\frac{d\theta}{dt} = \alpha$$

with $\alpha = (\frac{1}{2}, 1)$, $\alpha = (2, 5)$ and $\alpha = (1, 2)$.

Exercise 1.6.6 Notice that in Exercise 1.6.5 when $\alpha = (2, 5)$ the orbit O wraps three times around $\mathbb{T}^2$ in the horizontal direction and five time in the vertical one forming a loop without self-intersections. Represent now the Klein bottle K by a unit square identifying the horizontal lines and the vertical ones in the proper way.

Can you draw on K a loop without self-intersection having the same property as the orbit O described above? Motivate the answer.

Exercise 1.6.7 Consider a linear flow

$$\frac{d\theta}{dt} = \alpha$$

where $\alpha = (2, 5)$ on $\mathbb{T}^2$. In the previous chapter we consider the section $S = \{\theta_2 = 0\}$ which is a loop. Find another loop which is a section of the torus and that is everywhere transversal to the flow. Describe the first return map of to the new section. How many such sections are there?

Chapter 2
Topology of Circle Homeomorphisms

2.1 General Definitions

We collect in the following a list of basic definitions in dynamics.

A dynamical system is a map $f : X \to X$ where X is a metric space called the "state space" of the system. Starting with a state $x \in X$, f describes how the state changes with the time. Namely after n seconds the state x becomes $x_n = f^n(x)$, where the iteration f^n is defined by

$$f^n = f \circ f^{n-1} \text{ and } f^0 = id_X.$$

Given $f : X \to X$, the orbit of a point $x \in X$ is defined as

$$O(x) = \{f^n(x)\}_{n\in\mathbb{N}}.$$

We say that x is a fixed point for f if $f(x) = x$. The point $x \in X$ is a periodic point if there exists $n \in \mathbb{N}$ such that $f^n(x) = x$. The period of x is the smallest n such that $f^n(x) = x$.

The limit set $\omega(x)$ of x is defined as

$$\omega(x) = \left\{y \in X : \exists t_n \in \mathbb{N}, t_n \to +\infty \text{ with } f^{t_n}(x) \to y\right\}.$$

The limit set $\omega(x)$ is the set of accumulation of the orbit of x.

Fix a measure λ in the metric space X. We say that a set $A \subset X$ is an attractor for the system f if

$$\lambda\left(\{x \in X | A = \omega(x)\}\right) > 0.$$

Y. Dong et al., *One-Dimensional Dynamics*, Lecture Notes of the Unione Matematica Italiana 30, https://doi.org/10.1007/978-3-031-94882-4_2

Note that the definition of the limit set $\omega(x)$ is equivalent to

$$\omega(x) = \bigcap_{n\geq 1} \overline{O(f^n(x))},$$

therefore the definition of attractor presented here is equivalent to the definition in (1.2.1).

A nonempty set $A \subset X$ is minimal if for all $x \in A$, $\omega(x) = A$. All orbits are dense in a minimal set A.

We say that a measure μ on X is invariant under f if f is measurable and for all Borel set $A \subset X$,

$$\mu(f^{-1}(A)) = \mu(A).$$

For a precise definition of measurable function and Borel set, see [Pet83]. When f is continuous the previous definition is equivalent to ask that for all open set $U \subset X$, $\mu(f^{-1}(U)) = \mu(U)$.

The importance of the concept of invariant measure will be discussed in detail in later chapters.

2.1.1 Definition and Properties of Circle Homeomorphisms

The simplest non-trivial examples of dynamical systems are homeomorphisms of the circle which we define in the following.

We denote by $\mathbb{R}$ the real line and we denote by $\sim$ the equivalence relation on $\mathbb{R}$ defined by:

$$x \sim y \text{ if } x - y \in \mathbb{Z}.$$

We denote the equivalence class of $x \in \mathbb{R}$ by $[x] := \{y \in \mathbb{R} \,|\, y \sim x\}$. The equivalence class of 0 is then just $\mathbb{Z}$. Any equivalence class is a translation of $\mathbb{Z}$ by an element of the class, namely $[x] = x + \mathbb{Z}$.

We define the circle $\mathbb{S}^1$ as the set of all equivalence classes

$$\mathbb{S}^1 = \mathbb{R}/\mathbb{Z} := \{x + \mathbb{Z} | x \in \mathbb{R}\}.$$

Observe that every equivalence class has a unique element in the interval $[0, 1)$, you can then think of $\mathbb{S}^1$ as the interval $[0, 1]$ where 0 and 1 are identified, glued together.

Consider the natural projection π of $\mathbb{R}$ on $\mathbb{S}^1$ as

$$\pi : x \longrightarrow x + \mathbb{Z}.$$

Observe that π is periodic, i.e. for $x \in \mathbb{R}$, $\pi(x+1) = \pi(x)$. Moreover the natural projection allows us to "lift" any continuous function of the circle to a function on $\mathbb{R}$. More precisely:

Proposition 2.1.1 *If $f : \mathbb{S}^1 \longrightarrow \mathbb{S}^1$ is continuous, then there exists a continuous function $F : \mathbb{R} \longrightarrow \mathbb{R}$ such that*

$$f \circ \pi = \pi \circ F.$$

The map F is called a lift of f to $\mathbb{R}$ and it is unique up to an integer additive constant.

Observe that if f is an endomorphism of $\mathbb{S}^1$, then its lift $F : \mathbb{R} \longrightarrow \mathbb{R}$ has the same properties of continuity and differentiability of f.

Proposition 2.1.2 *Let $f : \mathbb{S}^1 \longrightarrow \mathbb{S}^1$ be a continuous function and let F be its lift. Then the quantity*

$$\deg(f) := F(x+1) - F(x)$$

is an integer independent of $x \in \mathbb{R}$. It is called the degree of f.

Definition 2.1.3 Let $f : \mathbb{S}^1 \longrightarrow \mathbb{S}^1$ be a continuous function and let F be its lift. We say that f is orientation preserving if F is a strictly increasing function. Similarly, we say that f is orientation reversing if F is a strictly decreasing function.

Lemma 2.1.4 *A circle homeomorphism has degree 1 (resp. -1) if it is orientation preserving (resp. reversing).*

The proofs of Propositions 2.1.1, 2.1.2, and Lemma 2.1.4 are left as exercises. For orientation reversing circle homeomorphisms the dynamics are simple (see Exercise 2.5.7). The whole course will be devoted to the study of orientation preserving circle homeomorphisms.

Remark 2.1.5 Let $f : \mathbb{S}^1 \longrightarrow \mathbb{S}^1$ be an orientation preserving circle homeomorphism, then by the definition we can suppose that any lift $F : \mathbb{R} \longrightarrow \mathbb{R}$ is continuous and has the following properties:

1. for all $x \in R$, $F(x+1) = F(x) + 1$,
2. F is a strictly increasing function of $\mathbb{R}$, namely, $F(x_1) < F(x_2)$ if $x_1 < x_2$.
3. For all $k \in \mathbb{N}$ the function F^k is strictly increasing,
4. for all $x \in \mathbb{R}$, $k \in \mathbb{N}$, $r \in \mathbb{Z}$, $F^k(x+r) = F^k(x) + r$.

In the following we give another representation of orientation preserving circle homeomorphisms which will be frequently used throughout the course and is based on the idea of "cutting" the circle, see Fig. 2.1. Let f be an orientation preserving circle homeomorphism and let F be the lift of f with the property that $F(0) = v \in [0, 1)$. Let $c = F^{-1}(1)$. Given f we define two maps f_- and f_+, and an interval map $\tilde{f} : [0, 1) \to [0, 1)$ by

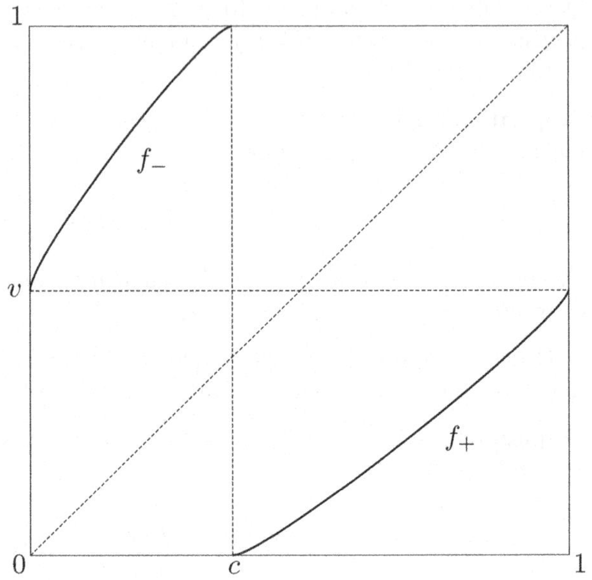

Fig. 2.1 Interval representation of a circle homeomorphism

$$\tilde{f}(x) = \begin{cases} f_-(x) = F(x) & x \in [0, c) \\ f_+(x) = F(x) - 1 & x \in [c, 1) \end{cases} \tag{2.1.1}$$

The point c is called the cutting point of the map. Observe that $f_- : [0, c) \to [v, 1)$ and $f_+ : [c, 1) \to [0, v)$ are increasing homeomorphisms.

Lemma 2.1.6 *Let f be an orientation preserving circle homeomorphism, let F be the lift with the property $F(0) \in [0, 1)$ and let $\tilde{f} : [0, 1) \to [0, 1)$ be the corresponding interval map given by formula (2.1.1). Define $\varphi : [0, 1) \to \mathbb{R}$ as*

$$\varphi(x) = \begin{cases} 0 & x \in [0, c) \\ 1 & x \in [c, 1) \end{cases}$$

Then for every $x \in [0, 1)$ and for every $n \in \mathbb{N}$,

$$F^n(x) = \tilde{f}^n(x) + \sum_{i=0}^{n-1} \varphi(\tilde{f}^i(x)).$$

Proof The proof goes by induction on n. For the case $n = 1$ observe that from the definition of $\tilde{f}$, (2.1.1), $F(x) = \tilde{f}(x) + \varphi(x)$. We suppose now that the lemma is true for n and we prove it for $n + 1$. Using the induction hypothesis and point 4. of Remark 2.1.5, we have

$$
\begin{aligned}
F^{n+1}(x) &= F(\tilde{f}^n(x) + \sum_{i=0}^{n-1} \varphi(\tilde{f}^i(x))) \\
&= F(\tilde{f}^n(x)) + \sum_{i=0}^{n-1} \varphi(\tilde{f}^i(x)) \\
&= \tilde{f}^{n+1}(x) + \varphi(\tilde{f}^n(x)) + \sum_{i=0}^{n-1} \varphi(\tilde{f}^i(x)) \\
&= \tilde{f}^{n+1}(x) + \sum_{i=0}^{n} \varphi(\tilde{f}^i(x)).
\end{aligned}
$$

□

The simplest examples of orientation preserving circle homeomorphisms, which will be constantly used throughout the sequel, are the rotations. Let $\rho \in [0, 1)$, using the previous characterization, the rotation of angle ρ, $r_\rho : [0, 1) \to [0, 1)$ is defined as

$$
r_\rho(x) = x + \rho \bmod 1.
$$

Notice that in the case of the rotations, the corresponding lift F and $\tilde{f}$ have constant derivative 1. This is the characterizing property of the rotations.

Note that $f(x) = \tilde{f}(x)$ for all $x \in [0, 1)$. For the simplicity of notation, in future discussions we will identify the interval map $\tilde{f}$ with its corresponding circle homeomorphism f.

2.2 Rotation Number

In 1885 Poincaré introduced the concept of rotation number for orientation preserving circle homeomorphisms, see [Poi93]. The rotation number of a circle map is the asymptotic average of the rotation of a point on the circle under the action of the map. In this section we introduce the concept of rotation number and we give some of its fundamental properties.

Throughout the section, f will be an orientation preserving homeomorphism of the circle and F its lift.

Proposition 2.2.1 *The quantity,*

$$
\tau(F) := \lim_{n\to\infty} \frac{1}{n}(F^n(x) - x) \tag{2.2.1}
$$

exists for all $x \in \mathbb{R}$ and it is independent of x. It depends from the choice of the lift in the following way: if $\tilde{F}$ is another lift of f, then $\tau(F) - \tau(\tilde{F}) \in \mathbb{Z}$.

We call $\tau(F)$ *the* translation number *of* F.

Observe that $\lim_{n\to\infty}\frac{1}{n}(F^n(x)-x)=\lim_{n\to\infty}\frac{1}{n}\sum_{i=0}^{n-1}F(F^i(x))-F^i(x)$, so the translation number is giving the average jump along the orbit of x.

For proving Proposition 2.2.1 we need the following lemmas:

Lemma 2.2.2 *For every* $n\in\mathbb{N}$, *the nth jump function* $J_n : x\to F^n(x)-x$ *has the following property: for every* $x, y\in\mathbb{R}$,

$$|J_n(x)-J_n(y)|\le 2. \tag{2.2.2}$$

Proof By point 4. of Remark 2.1.5, J_n is periodic of period 1, namely,

$$J_n(x+1)=F^n(x+1)-x-1=F^n(x)-x=J_n(x).$$

Hence it suffices to prove (2.2.2) for every $x, y\in[0,1]$. In fact directly from the definition of the Jump function and point 3. of Remark 2.1.5 we get

$$|J_n(x)-J_n(y)|\le|F^n(x)-F^n(y)|+|x-y|\le|F^n(1)-F^n(0)|+1.$$

Notice that by point 4. of Remark 2.1.5, $F^n(1)=F^n(0)+1$. Consequently

$$|J_n(x)-J_n(y)|\le 2.$$

□

The following lemma describes a classical property of sub-additive sequences.

Lemma 2.2.3 *Let* $(a_n)_{n\in\mathbb{N}}$ *be a sequence of real numbers satisfying the sub-additivity condition, i.e. there exists* $L\in\mathbb{R}$ *such that*

$$a_{m+n}\le a_n+a_m+L,$$

for all $m, n\in\mathbb{N}$. *Then the limit*

$$\lim_{n\to\infty}\frac{a_n}{n}\in\mathbb{R}\cup\{-\infty\}$$

exists.

Proof Let us consider the limit inferior $a:=\underline{\lim}_{n\to\infty}a_n/n\in\mathbb{R}\cup\{-\infty\}$. First observe that

$$a_{2^n}\le 2a_{2^{n-1}}+L.$$

Hence,

$$\frac{a_{2^n}}{2^n}\le\frac{a_{2^{n-1}}}{2^{n-1}}+\frac{L}{2^n}\le\frac{a_{2^{n-2}}}{2^{n-2}}+\frac{L}{2^{n-1}}+\frac{L}{2^n}\le\cdots\le\frac{a_2}{2}+L\sum_{k=2}^{n}\frac{1}{2^k}.$$

Then $a_{2^n}/2^n \le a_2/2 + L$ for all $n \in \mathbb{N}$, hence the limit inferior a is bounded from above by $a_2/2 + L$.

Next we show that $\overline{\lim}_{n\to\infty} a_n/n \le c$ for arbitrary $c > a$. This would imply that the limit exists and it is equal to a. For every $c > a$, let $b \in \mathbb{R}$ such that $a < b < c$. Since a is the limit inferior, we can find $n_0 \in \mathbb{N}$ larger than $\max(b, 4L/(c-b))$ such that $a_{n_0}/n_0 < b$. Let $n \ge \max(n_0, 2(\max_{r<n_0} a_r)/(c-b))$. We can write $n = n_0 k + r$ with $k, r \in \mathbb{N}$ and $r < n_0$, hence by the sub-additivity condition,

$$\frac{a_n}{n} \le \frac{k a_{n_0} + a_r + (k+1)L}{n} \le \frac{a_{n_0}}{n_0} + \frac{a_r}{n} + \frac{2L}{n_0}$$
$$< b + a_r \cdot \frac{c-b}{2\max_{r<n_0} a_r} + 2L \cdot \frac{c-b}{4L} \le c.$$

□

We are now ready to prove Proposition 2.2.1.

Proof of Proposition 2.2.1 We first prove the existence and independence in x for $\tau(F)$ when the lift F satisfies $F(0) \in [0, 1)$.

Existence. For any $x \in \mathbb{R}$, notice that $\{J_n(x)\}$ satisfies the sub-additivity condition, indeed

$$\begin{aligned} J_{m+n}(x) &= F^{m+n}(x) - F^n(x) + F^n(x) - x \\ &= J_m(F^n(x)) + J_n(x) \\ &= J_m(F^n(x)) - J_m(x) + J_m(x) + J_n(x). \end{aligned}$$

Since $J_m(F^n(x)) - J_m(x) \le 2$ by Lemma 2.2.2, we have $J_{m+n}(x) \le 2 + J_m(x) + J_n(x)$. By Lemma 2.2.3 the sequence $J_n(x)/n$ is convergent to a number $\tau(F) \in \mathbb{R} \cup \{-\infty\}$. Moreover by Lemma 2.1.6 we have

$$\tau(F) = \lim_{n\to\infty} \frac{\tilde{f}^n(x) + \sum_{i=0}^{n-1} \varphi(\tilde{f}^i(x)) - x}{n} \ge \lim_{n\to\infty} \frac{0 + n \cdot 0 - x}{n} = 0.$$

This implies that $\tau(F)$ is a finite number.

Independence from x. Let $x, y \in \mathbb{R}$. By Lemma 2.2.2,

$$\lim_{n\to\infty} \frac{1}{n}(|J_n(x) - J_n(y)|) \le \lim_{n\to\infty} \frac{2}{n} = 0.$$

Hence

$$\left| \lim_{n\to\infty} \frac{1}{n}(F^n(x) - x) - \lim_{n\to\infty} \frac{1}{n}(F^n(y) - y) \right| = \left| \lim_{n\to\infty} \frac{1}{n}(J_n(x) - J_n(y)) \right| = 0.$$

The translation number is then independent of the choice of the point x.

Moreover, for another lift $\widetilde{F} = F + m$ with $m \in \mathbb{Z}$, we have

$$\tau(\widetilde{F}) = \lim_{n\to\infty} \frac{1}{n}(F^n(x) + nm - x) = \tau(F) + m.$$

This explains the dependence of $\tau(F)$ on the choice of lift, and implies the existence and independence on x of $\tau(F)$ for arbitrary lift F. □

The proposition 2.2.1 justify the following terminology:

Definition 2.2.4 The *rotation number* $\rho(f)$ of f is defined as $\tau(F)$ modulo 1.

2.2.1 Properties of the Rotation Number

Proposition 2.2.5 *If $\rho(f) = 0$, then f has a fixed point.*

The proof is left as an exercise, see Exercise 2.5.8.

Proposition 2.2.6 *$\rho(f)$ is rational if and only if f has a periodic point.*

Proof Suppose that f has a periodic point $[x] \in \mathbb{S}^1$ of period $q \in \mathbb{N}$, namely $f^q([x]) = [x]$. Equivalently there exists $p \in \mathbb{Z}$ such that for a chosen lift F, $F^q(x) = x + p$. Then, by point 4. of Remark 2.1.5, if $m \in \mathbb{N}$

$$\frac{F^{mq}(x) - x}{mq} = \frac{x + mp - x}{mq} = \frac{p}{q}$$

and then $\rho(f) = \frac{p}{q}$.

For proving the other implication we observe that, by the definition of translation number

$$\tau(F^m) = \lim_{n\to\infty} \frac{1}{n}((F^m)^n(x) - x) = m \lim_{n\to\infty} \frac{1}{mn}(F^{mn}(x) - x) = m\tau(F). \tag{2.2.3}$$

If $\rho(f) = \frac{p}{q} \in \mathbb{Q}$, note that F^q is a lift of f^q, using (2.2.3) we have

$$\rho(f^q) = \tau(F^q) \mod 1 = q \cdot \frac{p}{q} \mod 1 = 0.$$

By Proposition 2.2.5, f has a periodic point. □

Proposition 2.2.7 *Let ρ be an irrational number and let r_ρ be the rotation of angle ρ. Then, for every $x \in \mathbb{S}^1$, $\omega(x) = \mathbb{S}^1$, namely all orbits are dense, $\mathbb{S}^1$ is a minimal set.*

Proof Notice that, by the definition of rotation, the nth iteration r_ρ^n is a rotation over angle $n\rho$ mod 1. Given $\epsilon > 0$ we are going to prove that there exists $n \in \mathbb{N}$

such that $0 < n\rho \bmod 1 < \epsilon$. Let $z \in \omega(0)$. Then there exist $s < t$ such that $r_\rho^s(0), r_\rho^t(0) \in (z - \frac{\epsilon}{2}, z + \frac{\epsilon}{2})$. Being ρ irrational, by Proposition 2.2.6, $r_\rho^s(0) \neq r_\rho^t(0)$.Let $n := t - s$, then $r_\rho^n(r_\rho^s(0)) = r_\rho^t(0)$, hence r_ρ^n is a rotation of angle smaller than $|r_\rho^t(0) - r_\rho^s(0)| \leq \epsilon$.

Fix $x \in \mathbb{S}^1$. Let $I \subset \mathbb{S}^1$ be an open interval of length $\epsilon > 0$. We will show that the orbit of x under r_ρ intersects I. This would imply that the orbit of x is dense in $\mathbb{S}^1$. By the previous argument there exists $n \in \mathbb{N}$ such that r_ρ^n is a rotation of angle strictly smaller than ϵ. To simplify the notation let r denote r_ρ^n. For all $j \geq 0$ we define the interval $I_j = [r^j(x), r^{j+1}(x)]$. Observe that all such intervals have size smaller than ϵ and that they are adjacent, namely $I_j \cap I_{j+1} = r^{j+1}(x)$. Hence,

$$\mathbb{S}^1 = \bigcup_{j \geq 0} I_j.$$

Therefore there exists $j \in \mathbb{N}$ such that $I_j \cap I \neq \emptyset$. Observe that the boundary points of the intervals I_j are iterates of x. Because of the fact that $|I| = \epsilon > |I_j|$, I contains at least one of the boundary points of I_j. It contains at least one point of the orbit of x. □

2.3 Equivalent Systems

In the introduction we discussed the notion of equivalent systems. Two systems are equivalent if they have the same dynamical properties. For example, if a system has three periodic orbits of period 5 then any equivalent system should also have three periodic orbit of period 5. If the system has an invariant Cantor set then any equivalent system should also have an invariant Cantor set. If a system is obtained from another system by just a change of coordinates then the two systems are equivalent. If one understands the first systems then clearly the second one is also understood. The definition of equivalent systems is based on the idea of a coordinate change. The coordinate change in the general context is called a conjugacy. Equivalent systems are called conjugate.

Definition 2.3.1 Given two dynamical systems $f : X \to X$ and $g : Y \to Y$. We say that f and g are conjugate, topologically equivalent, if there exists a homeomorphism $h : X \to Y$ such that

$$h \circ f = g \circ h.$$

The map h is called a conjugacy.

Remark 2.3.2 Here are some general remarks concerning conjugacies.

1. Observe, a conjugacy maps orbits to orbits. Namely,

$$h(O(x)) = O(h(x)).$$

That means that every orbit of f has a corresponding orbit for g.
2. A conjugacy sends periodic orbits into periodic orbits of the same period.
3. Conjugacy defines an equivalence relation between systems. This allows us to speak about (topological) equivalence classes of systems, see the introduction.
4. Generally, the conjugacy between equivalent systems is not unique.

Definition 2.3.3 Given two dynamical systems $f : X \to X$ and $g : Y \to Y$. We say that f is semi-conjugate to g if there exists a continuous surjective map $h : X \to Y$ such that

$$h \circ f = g \circ h.$$

If f is semi-conjugate to g then f has more orbits than g. Namely, for every $y \in Y$ and $h(x) = y$ we have $O(y) = h(O(x))$. The dynamics of f is richer than the dynamics of g.

The next proposition tells that the rotation number is a topological invariant: equivalent systems have the same rotation number. The proof is left as an exercise, see Exercise 2.5.11.

Proposition 2.3.4 *Let f and g be orientation preserving circle homeomorphisms, if f is semi-conjugate to g, then $\rho(f) = \rho(g)$.*

The previous proposition and the following theorem summarize the classical theory of Poincaré on circle homeomorphisms.

Theorem 2.3.5 *Let f be a orientation preserving circle homeomorphism with irrational rotation number ρ. Then f is semi-conjugate to the rotation over angle ρ.*

Proof For all $n \in \mathbb{N}$ we define a measure

$$\mu_n = \frac{1}{n} \sum_{i=0}^{n-1} \delta_{f^i(0)}$$

where for all i, $\delta_{f^i(0)}$ is the point mass concentrated in $f^i(0)$. Notice that, for all $n \in \mathbb{N}$, μ_n is a probability measure on the circle and by the compactness of the circle we can take a weak limit μ of μ_n. Namely, there exists a sequence $\{k_n\}_{n \in \mathbb{N}}$, $k_n \to \infty$ such that $\mu_{k_n} \to \mu$. Even if it is not formally correct, for simplifying the notation we will assume that for all $n \in \mathbb{N}$ $k_n = n$, that is, $\mu_n \to \mu$.

Define $h : \mathbb{S}^1 \to \mathbb{S}^1$ and $r : \mathbb{S}^1 \to \mathbb{S}^1$ by

$$h(x) = \int_0^x d\mu,$$
$$r(x) = h \circ f \circ h^{-1}(x).$$

We will show that h is a semi-conjugacy, and r is indeed the rotation of angle ρ. First we prove the following claim:

- Claim 1. μ is invariant under f.
- Claim 2. If $J \subset \mathbb{S}^1$ is any interval, then $|h(J)| = \mu(J)$, where $|\cdot|$ is the diameter.

Proof of the Claim Let $U \subset \mathbb{S}^1$ be an open set, then

$$\begin{aligned}\mu(f^{-1}(U)) &= \lim_{n\to\infty} \frac{1}{n}\sum_{i=0}^{n-1} \delta_{f^i(0)}(f^{-1}(U)) \\ &= \lim_{n\to\infty} \frac{1}{n}\sum_{i=0}^{n-1} \delta_{f^{i+1}(0)}(U) \\ &= \lim_{n\to\infty} \frac{1}{n}\sum_{i=1}^{n} \delta_{f^i(0)}(U) \\ &= \lim_{n\to\infty} \left[\mu_n(U) + \frac{1}{n}\delta_{f^n(0)}(U) - \frac{1}{n}\delta_0(U)\right] = \mu(U).\end{aligned}$$

This proves Claim 1. Claim 2 follows directly from the definition of h. ■

(i) h is continuous. We prove that for every $x \in \mathbb{S}^1$,

$$\lim_{\ell\to 0} \mu([x-\ell, x+\ell]) = 0. \tag{2.3.1}$$

By contradiction we assume that there exist $x \in \mathbb{S}^1$ and a sequence $\{\ell_n\}_{n\in\mathbb{N}}$, $\ell_n \to 0$ such that $\mu([x-\ell_n, x+\ell_n]) \geq a$ with $a > 0$. Consider the points $x_i = f^i(x)$ for $i = \{0, 1, \ldots, \lfloor 1/a \rfloor + 1\}$. The hypothesis $\rho \notin \mathbb{Q}$, implies that $x_i \neq x_j$ for $i \neq j$. We define $L := \min_{i\neq j} |x_i - x_j|$ as the minimal distance between the points x_i and let

$$J_i^n = f^i\left([x-\ell_n, x+\ell_n]\right).$$

Because f is continuous there exists $n_0 \in \mathbb{N}$ such that, for all $i = \{0, 1, \ldots, \lfloor 1/a \rfloor + 1\}$, $|J_i^{n_0}| < \frac{1}{2}L$. Consequently $J_i^{n_0}$ are pairwise disjoint, i.e. for all $i, j \in \{0, 1, \ldots, \lfloor 1/a \rfloor + 1\}$, $i \neq j$, $J_i^{n_0} \cap J_j^{n_0} = \emptyset$. By Claim 1, $\mu(J_i^{n_0}) = \mu(J_0^{n_0}) \geq a$ for all i, then

$$1 \geq \sum_{i=0}^{\lfloor \frac{1}{a} \rfloor + 1} \mu(J_i^{n_0}) \geq \left(\left\lfloor \frac{1}{a} \right\rfloor + 1\right) a > 1.$$

This contradiction proves (2.3.1). Hence h is continuous.

(ii) h is surjective. From the definition of h, it is clear that $h(0) = 0$ and $\lim_{x\uparrow 1} h(x) = 1$. Since we already proved the continuity of h we have that h is surjective.

(iii) r is well-defined. Observe that h is increasing, then for all $x \in \mathbb{S}^1$ $h^{-1}(x)$ is either a point or an interval. In the first case, $r(x)$ is well defined. In the second one, consider the interval $J = h^{-1}(x)$ and observe that $\mu(J) = 0$. Hence by Claim 2, $|r(x)| = |h(f(J))| = \mu(f(J))$. By Claim 1 we have $\mu(f(J)) = \mu(J) = 0$. That is, $|r(x)| = 0$, $r(x)$ is a point. Hence r is well-defined.

(iv) r is the rotation of angle ρ.Use the characterizing property of the rotation, it suffices to show that r has constant derivative 1. For proving that, we chose any interval $J \subset \mathbb{S}^1$ and we prove that $|r(J)| = |J|$. Indeed, using Claim 1 and 2, we have

$$|r(J)| = |h(f(h^{-1}(J)))| = \mu(f(h^{-1}(J))) = \mu(h^{-1}(J)) = |h(h^{-1}(J))| = |J|.$$

We finally proved that r is a rotation. Since f is semi-conjugate to r, by Proposition 2.3.4, r has angle ρ.

□

Remark 2.3.6 Notice that the semi-conjugacy constructed, being the integral of a measure, is increasing. In fact, any semi-conjugacy between circle homeomorphisms with irrational rotation number is increasing (Exercise 2.5.12). Hence in the following discussion without loss of generality we will assume that a semi-conjugacy between circle maps is increasing.

In Chap. 3 we will discuss the conjugation problem for circle diffeomorphisms. It will be shown that, indeed, the semi-conjugacy from Theorem 2.3.5 is in fact a conjugacy, a homeomorphism.

2.4 Full Families

In the following we define a "family" as a function $\mathcal{F} : [0, 1) \times [0, 1) \to [0, 1)$ continuous in t, such that for every $t \in [0, 1)$ the map $f_t : [0, 1) \to [0, 1)$ is a circle homeomorphism, where, $\forall x \in [0, 1)$,

$$f_t(x) := \mathcal{F}(x, t).$$

A family is said to be "full" if for every $\rho \in [0, 1)$ there exists $t \in [0, 1)$ such that the rotation number $\rho(f_t) = \rho$.

Theorem 2.4.1 *Given a family $\mathcal{F} : [0, 1) \times [0, 1) \to [0, 1)$, if $\rho(f_0) = 0$ and $\lim_{t\to 1} \rho(f_t) = 1$, then $\mathcal{F}$ is a full family.*

In the rest of the section we will give two different proofs for Theorem 2.4.1. The first one is classical in this context, while the second one uses a method which can be applied in much more general situations.

2.4.1 Classical Method

We start studying the dependence of the rotation number on the function. Then the proof of Theorem 2.4.1 is a direct application of Proposition 2.4.3 and the intermediate value theorem.

Remark 2.4.2 Composition as an operator is continuous in the uniform topology. In particular, given any fixed function f, for every $\varepsilon > 0$ and $q \in \mathbb{N}$ there exists $\delta > 0$ such that if $d(f, g) < \delta$ then $d(f^q, g^q) < \varepsilon$, where $d(f, g)$ is the metric induced by the uniform norm of the lifts, defined as following: $d(f, g) = \sup_x |F(x) - G(x)|$, where F, G are lifts of f, g respectively, satisfying $|F(0) - G(0)| \leq 1/2$.

Proposition 2.4.3 *$f \mapsto \rho(f)$ is continuous in the uniform topology.*

Proof We fix an orientation preserving homeomorphism f, with $\rho(f) = \rho$. Let $r, r' \in \mathbb{Q}$ such that $r' < \rho < r$. Fix a lift F of f and $x \in \mathbb{R}$. By the definition of translation number, there exist $p, q \in \mathbb{N}$ such that

$$\frac{F^q(x) - x}{q} < \frac{p}{q} = r.$$

This implies that, $F^q(x) < x + p$. By the continuity of iteration, we get that if a circle map $\tilde{f}$ is close enough to f, then it has a lift $\tilde{F}$ such that

$$\tilde{F}^q(x) < x + p.$$

Consequently, for all $m \in \mathbb{N}$, by point 4. of Remark 2.1.5 we have

$$\begin{aligned}
\tilde{F}^{mq}(x) &= \tilde{F}^{(m-1)q}(\tilde{F}^q(x)) \\
&\leq \tilde{F}^{(m-1)q}(x + p) \\
&= \tilde{F}^{(m-1)q}(x) + p \\
&\leq \dots \\
&= x + mp.
\end{aligned}$$

Hence, the translation number of $\tilde{F}$,

$$\tau(\tilde{F}) = \lim_{m\to\infty} \frac{\tilde{F}^{mq}(x) - x}{mq} \leq \frac{p}{q} = r$$

and $\rho(\tilde{f}) \leq r$. Similarly we prove that when $\tilde{f}$ is close enough to f, then $\rho(\tilde{f}) \geq r'$. Finally the rotation number is continuous. □

In the following we discuss the properties of monotonicity of the rotation number. Notice that by Definition 2.2.1 the translation number is a monotone function. In fact, if for all $x \in \mathbb{R}$, $F_1(x) \geq F_2(x)$ then $\tau(F_1) \geq \tau(F_2)$. This brings us naturally to introduce the concept of order in the circle.

Definition 2.4.4 We define a partial order "$\prec$" on the collection of orientation preserving circle homeomorphisms that we are studying as follows. Let F_0, F_1 be the respective lifts of circle homeomorphisms f_0, f_1 satisfying $F_0(0), F_1(0) \in [0, 1)$. We write $f_0 \prec f_1$ if all $x \in \mathbb{R}$ but those that belong to finitely many equivalent classes of $\mathbb{R}/\mathbb{Z}$, $F_0(x) < F_1(x)$.

The definition of rotation number imply immediately the following proposition:

Proposition 2.4.5 *The map $\rho(\cdot)$ is increasing: if $f_0 \prec f_1$, then $\rho(f_0) \leq \rho(f_1)$.*

Remark 2.4.6 In particular, if $(f_t)_{t\in\mathbb{R}}$ is a family of orientation preserving circle homeomorphisms such that for each x, $f_t(x)$ is increasing in t, then $\rho(f_t)$ is increasing in t.

Moreover, for irrational values, the rotation number is strictly increasing (the proof is left as an exercise, see Exercise 2.5.17).

Proposition 2.4.7 *If $f_0 \prec f_1$ and if $\rho(f_0)$ is irrational, then $\rho(f_0) < \rho(f_1)$.*

Remark 2.4.8 The function $[0, 1] \ni t \to \rho(f_t) \in [0, 1]$ is a continuous function similar to the "Devil's Staircase": it's plateaux happen exactly at the rational values. If the family is strictly increasing then the rotation number is strictly increasing at the irrational values. For a discussion on the "Devil's Staircase", or the Cantor function see [Dov+06].

2.4.2 The Thurston Method

Topological equivalence says that there is a coordinate change, modifying one system into the other by conjugation. This means that they are topologically the same, that the orbits of the two systems have the same *orbit combinatorics*. In this section we are going to use the Thurston method to prove the Full Family Theorem. The method will also clarify the notion of orbit combinatorics.

In this subsection we prove again Theorem 2.4.1. The strategy of the proof is more general than the previous one and can be applied to many different settings. As you can notice reading the previous subsection, the classical proof of the full-family theorem is very specific in the sense that it uses the intermediate value theorem which is not available in higher dimensional parameter spaces. Moreover, the classical proof uses only one topological invariant: the rotation number. The proof that we present now is more general and applies also in contexts where several topological invariants play a role, with higher dimensional parameter spaces. The Brouwer fixed point Theorem replaces the intermediate value theorem.

For making the proof clearer we will use the representation of circle homeomorphisms presented in (2.1.1). Let $c : [0, 1) \to [0, 1)$ be a continuous function which will describe the cutting points of the maps in the family. Consider the continuous family $\mathcal{F} : [0, 1) \times [0, 1) \to [0, 1)$ of circle homeomorphisms defined, for all $x \in [0, 1)$ and $t \in [0, 1)$ as

$$\mathcal{F}(x, t) = f_t(x) = \begin{cases} f_-(x, t) & x \in [0, c(t)) \\ f_+(x, t) & x \in [c(t), 1) \end{cases} \tag{2.4.1}$$

where f_t is the circle homeomorphism defined by $f_t(x) = \mathcal{F}(x, t)$, for $x \in [0, 1)$, $f_-(x, t)$ and $f_+(x, t)$ are the corresponding branches f_- and f_+ of f_t as defined in (2.1.1). For simplifying the argument we will assume that the function $[0, 1) \ni t \to f_-(0, t) \in [0, 1)$ is an increasing homeomorphism. Then without loss of generality we may assume that for $t \in [0, 1)$

$$f_-(0, t) = t.$$

The following proposition is the core of the proof of Theorem 2.4.1 and for proving it we will use the so called "Thurston method".

Proposition 2.4.9 *For a family defined as above, for all* $\rho \in \mathbb{Q} \cap [0, 1)$ *there exists* $t \in [0, 1)$ *such that*

$$\rho(f_t) = \rho.$$

Before proving Proposition 2.4.9 we will give some examples showing concretely how the Thurston method works.

2.4.2.1 Examples of the Thurston Method

We consider the family as in Fig. 2.2, defined as,

$$F(x, t) = \begin{cases} f_-(x, t) = 2(1 - t)x + t & x \in [0, \frac{1}{2}) \\ f_+(x, t) = 2tx - t & x \in [\frac{1}{2}, 1) \end{cases} \tag{2.4.2}$$

Notice that the branches of the family, f_- and f_+ are affine and the cutting point is $c = \frac{1}{2}$. Indeed this family is slightly different from the general definition in (2.4.1), the cutting point here is fixed, and for $t = 0$ and $t = 1$, the map does not represent a circle homeomorphism. However this is the simplest family which allows to show the idea of the Thurston method.

As first example we fix $\rho = 1/3$ and we prove that there exists a parameter $t \in (0, 1)$ such that the rotation number of f_t, $\rho(f_t) = \frac{1}{3}$. Let us start studying the orbits of the rotation of angle $\frac{1}{3}$,

Fig. 2.2 The piecewise affine family

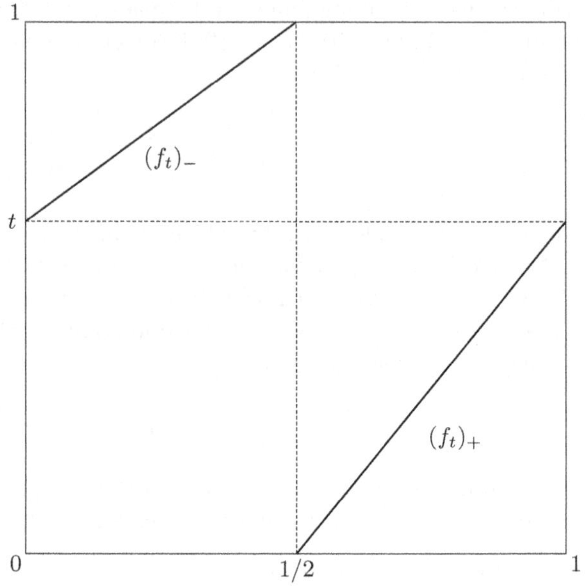

$$r_\rho(x) = \begin{cases} r_-(x) = x + \frac{1}{3} & x \in [0, c_\rho) \\ r_+(x) = x - \frac{2}{3} & x \in [c_\rho, 1) \end{cases}$$

Observe that in this case the cutting point is $c_\rho = \frac{2}{3} = 1 - \rho$. Let us consider now the combinatorial structure of the orbit of $\underline{v}_0 = 0$. We denote by

$$\underline{v}_i = r_\rho^i(\underline{v}_0) \text{ for } i = 1, 2. \tag{2.4.3}$$

Notice that $\underline{v}_0 = 0$, $\underline{v}_1 = \frac{1}{3}$, $\underline{v}_2 = \frac{2}{3} = c_\rho$. The orbit of 0, which is periodic, can then be dynamically represented as

$$0 = \underline{v}_0 \to \underline{v}_1 \to \underline{v}_2 = c_\rho \to \underline{v}_0 = 0 \tag{2.4.4}$$

and geometrically as

$$0 = \underline{v}_0 < \underline{v}_1 < \underline{v}_2 = c_\rho. \tag{2.4.5}$$

Observe that (2.4.5) contains two pieces of information. The first is the position of the points in the orbit, *orbit geometry*. Secondly, it describes the order of the points in the orbit. This is the *orbit combinatorics*. In the family there is most likely not a map such that the orbit geometry is exactly as given in (2.4.5). However, the goal is to find a map whose orbit of 0 has the same orbit combinatorics as (2.4.5). In particular, the goal is to find a parameter t such that the orbit of $v_0 = 0$ under f_t,

$$v_i = f_t^i(v_0) \text{ for } i = 1, 2,$$

satisfies the equivalent of (2.4.4) and (2.4.5), namely,

$$0 = v_0 \to v_1 \to v_2 = c \to v_0 = 0 \tag{2.4.6}$$

and

$$0 = v_0 < v_1 < v_2 = c. \tag{2.4.7}$$

Equations (2.4.6) and (2.4.7) describe the configurations associated with the orbit combinatorics given by (2.4.4) and (2.4.5). At this moment we don't know which specific configuration, satisfying equations (2.4.6) and (2.4.7), can be realized with a map in the family. Let us consider the whole space of a priori possible configurations. Namely, define the configurations space

$$C_{\frac{1}{3}} = \{(v_0, v_1, v_2) \in [0, 1)^3 | 0 = v_0 < v_1 < v_2 = c = \frac{1}{2}\}. \tag{2.4.8}$$

Observe that the configurations have only the right geometrical and combinatorial information given by (2.4.7). The configuration that we are looking for should also have the dynamical information given by (2.4.6). This configuration will be a fixed point of the Thurston map $T : C_{\frac{1}{3}} \to C_{\frac{1}{3}}$ which we are going to define now and it is illustrated in Fig. 2.3.

Given a configuration $v = (0, v_1, \frac{1}{2}) \in C_{\frac{1}{3}}$, there exists a unique $t \in (0, 1)$ such that

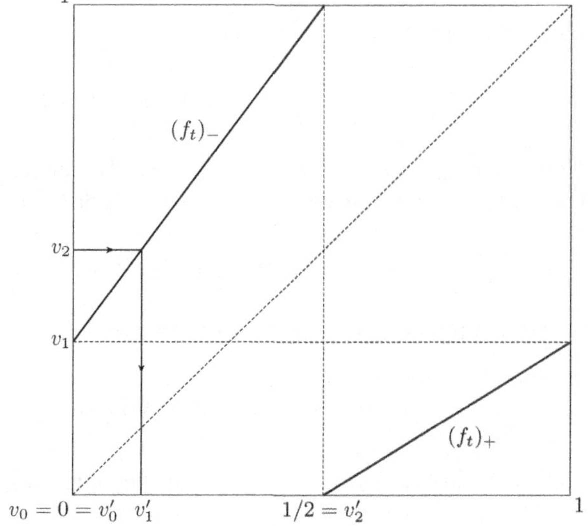

Fig. 2.3 The Thurston map for $\rho = 1/3$

$$f_t(v_0) = v_1. \tag{2.4.9}$$

Namely, $t = v_1$. Define $Tv = (0, v'_1, \frac{1}{2})$ such that

$$f_t(v'_1) = v_2 \tag{2.4.10}$$

or equivalently $f_-(v'_1, t) = \frac{1}{2}$. A calculation show that

$$v'_1 = \frac{\frac{1}{2} - v_1}{2(1 - v_1)}.$$

The action of T is "pulling back" the configuration using the preimage of f_t. A calculation shows that T has a fixed point $v^* = (0, v^*_1, v^*_2) = (0, t^*, \frac{1}{2}) = (0, \frac{3-\sqrt{5}}{4}, \frac{1}{2})$. We show that f_{t^*} has the following property:

$$f_{t^*} : 0 \to t^* \to \frac{1}{2} \to 0$$

which is equivalent to state that $\rho(f_{t^*}) = \frac{1}{3}$. First, notice that for all $t \in (0, 1)$, $f_t : 0 \to t$ and $f_t : \frac{1}{2} \to 0$. It remains to prove that $f_{t^*}(v^*_1) = \frac{1}{2}$ which is a direct consequence of the definition of the Thurston map (2.4.10) and the fact that v^* is a fixed point, $f_{t^*}(v^*_1) = v^*_2$.

As second example we fix $\rho = \frac{2}{3}$. Consequently the rotation of angle $\frac{2}{3}$ has the following form:

$$r_\rho(x) = \begin{cases} r_-(x) = x + \frac{2}{3} & x \in [0, c_\rho) \\ r_+(x) = x - \frac{1}{3} & x \in [c_\rho, 1) \end{cases}$$

with cutting point $c_\rho = \frac{1}{3}$. We denote by

$$\underline{v}_i = r^i_\rho(\underline{v}_0) \text{ for } i = 1, 2.$$

Notice that $\underline{v}_0 = 0$, $\underline{v}_1 = \frac{2}{3}$, $\underline{v}_2 = \frac{1}{3} = c_\rho$. Dynamically the orbit of 0 is as (2.4.4). Combinatorially and geometrically there is a difference:

$$0 = \underline{v}_0 < \underline{v}_2 = c_\rho < \underline{v}_1.$$

The configurations space for rotation number $\rho = \frac{2}{3}$ will be

$$C_{\frac{2}{3}} = \{(v_0, v_1, v_2) \in [0, 1)^3 | 0 = v_0 < v_2 = c = \frac{1}{2} < v_1\}$$

and the Thurston map $T : C_{\frac{2}{3}} \to C_{\frac{2}{3}}$ is defined as follows, see Fig. 2.4.

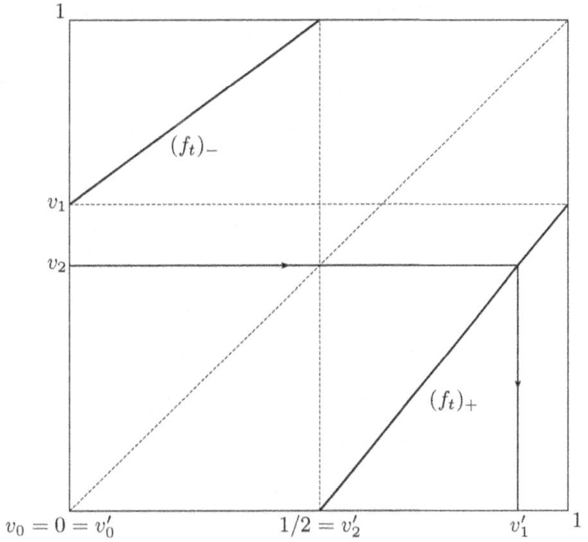

Fig. 2.4 The Thurston map for $\rho = 2/3$

Given a configuration $v = (0, v_1, \frac{1}{2}) \in C_{\frac{2}{3}}$ then there exists a unique $t \in (0, 1)$ such that

$$f_t(v_0) = v_1.$$

Namely, $t = v_1$. Consequently $Tv = (0, v_1', \frac{1}{2})$ where

$$f_t(v_1') = v_2$$

equivalently $f_+(v_1', t) = \frac{1}{2}$. Observe that the branch with which the Thurston map is pulling back is now changed because of the change of the combinatorics. In this case

$$v_1' = \frac{\frac{1}{2} + v_1}{2v_1},$$

and the fixed point for the Thurston map $v^* = (0, t^*, \frac{1}{2}) = (0, \frac{1+\sqrt{5}}{4}, \frac{1}{2})$. As before we can show that $\rho(f_{t^*}) = \frac{2}{3}$.

As third example we fix $\rho = \frac{2}{5}$. In this case the cutting point is $c_\rho = \frac{3}{5}$ We denote by

$$\underline{v}_i = r_\rho^i(\underline{v}_0) \text{ for } i = 1, 2, 3, 4.$$

Notice that $\underline{v}_0 = 0$, $\underline{v}_1 = \frac{2}{5}$, $\underline{v}_2 = \frac{4}{5}$, $\underline{v}_3 = \frac{1}{5}$ and $\underline{v}_4 = \frac{3}{5} = c_\rho$. As before dynamically the orbit of 0 is as

Fig. 2.5 The Thurston map for $\rho = 2/5$

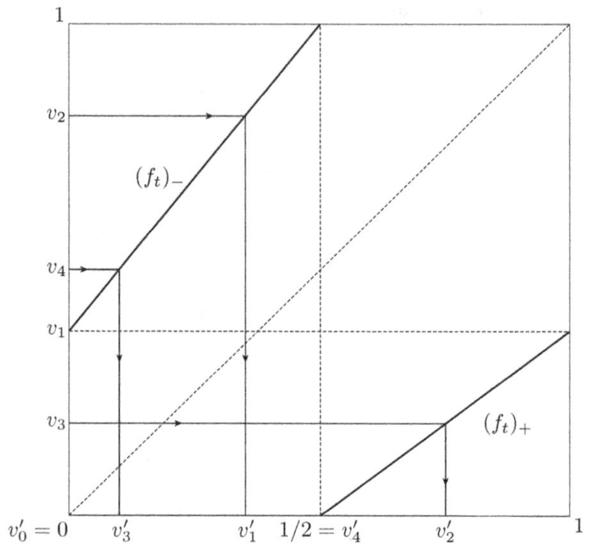

$$0 = \underline{v}_0 \to \underline{v}_1 \to \underline{v}_2 \to \underline{v}_3 \to \underline{v}_4 = c_\rho \to 0$$

but now combinatorially and geometrically we need

$$0 = \underline{v}_0 < \underline{v}_3 < \underline{v}_1 < \underline{v}_4 = c_\rho < \underline{v}_2. \tag{2.4.11}$$

The configurations space for rotation number $\rho = \frac{2}{5}$ will be

$$C_{\frac{2}{5}} = \{(v_0, v_1, v_2, v_3, v_4) \in [0,1)^5 | 0 = v_0 < v_3 < v_1 < v_4 = c < v_2\}$$

and the Thurston map $T : C_{\frac{2}{5}} \to C_{\frac{2}{5}}$ is defined as follows, see Fig. 2.5. Given a configuration $v = (0, v_1, v_2, v_3, v_4 = c) \in C_{\frac{2}{5}}$ then there exists a unique $t \in (0,1)$ such that

$$f_t(v_0) = v_1.$$

Namely, $t = v_1$. Define $Tv = (0, v'_1, v'_2, v'_3, c)$ where

$$f_t(v'_i) = v_{i+1} \text{ for } i = 1, 2, 3.$$

Observe that $f_-(v'_1, t) = v_2$, $f_+(v'_2, t) = v_3$, $f_-(v'_3, t) = v_4$, namely that the Thurston map is now pulling back using both branches according to the combinatorics.

To write down the precise formula for the fixed point of T, it is maybe possible, but surely very complicated. Instead we will use the Brouwer Fixed Point Theorem.

Notice that $C_{\frac{2}{5}}$ is a 3-dimensional simplex and T is continuous. Unfortunately $C_{\frac{2}{5}}$ is an open simplex, so we cannot directly apply the Brouwer Fixed Point Theorem. In this case and in general the main technical aspect is to extend T to the boundary of the configuration space and prove that the fixed point cannot be on the boundary. The general case is discussed in the next subsection.

As final remark on the last example we rewrite the configuration space $C_{\frac{2}{5}}$ in a form that can be easily generalized. Let $\underline{O}_{\frac{2}{5}} = \{0 = \underline{v}_0, \underline{v}_1, \underline{v}_2, \underline{v}_3, \underline{v}_4 = c_\rho\}$ the orbit of 0 under the rotation of angle $\frac{2}{5}$. Observe that the configuration space can be rewritten as follows,

$$C_{\frac{2}{5}} = \left\{\varphi : \underline{O}_{\frac{2}{5}} \to [0, 1) | \varphi \text{ is strictly increasing}, \varphi(\underline{v}_0) = 0\right\}.$$

In the following we will continue to use the convention that every $\varphi \in C_{\frac{2}{5}}$, is determined by the values. We identify φ with $(0, v_1, v_2, v_3, \frac{1}{2})$ such that $\varphi(\underline{v}_i) = v_i$.

2.4.2.2 General Discussion of the Thurston Method

In this section we discuss the Thurston Method for the general family given in (2.4.1). Fix a rational number $\rho = \frac{p}{q}$, $p, q \in \mathbb{N}$ co-prime. The rotation of angle $\frac{p}{q}$, $r_{\frac{p}{q}}$ is defined as

$$r_\rho(x) = \begin{cases} r_-(x) = x + \frac{p}{q} & x \in [0, c_\rho) \\ r_+(x) = x - 1 + \frac{p}{q} & x \in [c_\rho, 1) \end{cases}$$

where the cutting point $c_\rho = 1 - \frac{p}{q}$. As in the examples above we consider the periodic orbit of $0 = \underline{v}_0$ under $r_{\frac{p}{q}}$,

$$\underline{v}_i = r_\rho^i(\underline{v}_0) \text{ for } i = 1, 2, 3, q - 1.$$

Dynamically, the orbit of 0 is as

$$0 = \underline{v}_0 \to \underline{v}_1 \to \dots \underline{v}_{q-1} = c_\rho \to 0.$$

Combinatorially and geometrically we describe the orbit by the set

$$\underline{O}_{\frac{p}{q}} = \{\underline{v}_0, \underline{v}_1, \dots, \underline{v}_{q-1}\}.$$

In particular, $\underline{O}_{\frac{p}{q}}$ contains the combinatorial and geometrical information as in Eq. (2.4.11). The configuration space is

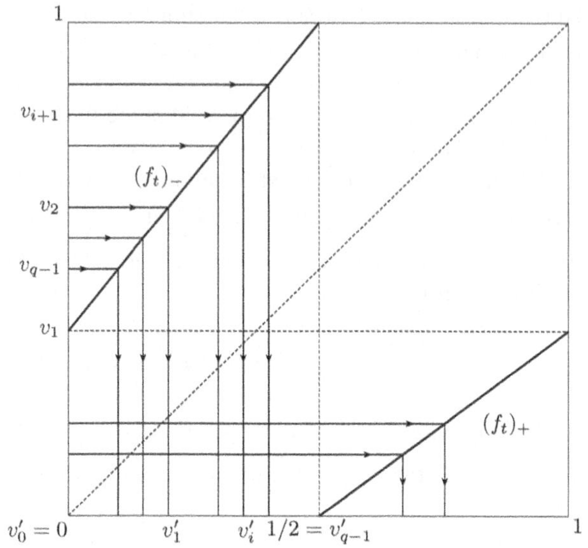

Fig. 2.6 The Thurston map for $\rho = p/q$

$$C_{\frac{p}{q}} = \left\{\varphi : \underline{O}_{\frac{p}{q}} \to [0,1) | \varphi \text{ is strictly increasing}, \varphi(\underline{v_0}) = 0\right\}.$$

The Thurston map $T : C_{\frac{p}{q}} \to C_{\frac{p}{q}}$ is defined as follows, see Fig. 2.6.

Given a configuration $v = (0, v_1, \ldots, v_{q-1}) \in C_{\frac{p}{q}}$, there exists a unique $t \in (0,1)$ such that

$$f_t(v_0) = v_1.$$

Namely, $t = v_1$. Define $Tv = (0, v'_1, \ldots, v'_{q-1})$ where

$$f_t(v'_i) = v_{i+1} \text{ for } i = 1, 2, \ldots, q-2,$$

and

$$f_t(v'_{q-1}) = v_0 = 0.$$

That is

$$v'_{q-1} = c(t).$$

Observe that the Thurston map is now pulling back using branches according to the combinatorics.

Lemma 2.4.10 *If* $v \in C_{\frac{p}{q}}$ *then* $Tv \in C_{\frac{p}{q}}$.

Proof Suppose $v = (0, v_1, \ldots, v_{q-1}) \in C_{\frac{p}{q}}$ and $Tv = (0, v'_1, \ldots, v'_{q-1})$. Let $t \in [0, 1]$ be such that $f_t(0) = v_1$. For all $i, j \in \{1, \ldots, q-1\}$ such that $\underline{v}_i < \underline{v}_j$, we want to show that $v'_i < v'_j$, which implies $Tv \in C_{\frac{p}{q}}$. We need to study three different cases.

I $\underline{v}_i < \underline{v}_j < c_\rho = \underline{v}_{q-1}$, i.e., $0, \underline{v}_i, \underline{v}_j$ are in the same branch (r_-) of r_ρ, hence

$$r_\rho(0) = \underline{v}_1 < r_\rho(\underline{v}_i) = \underline{v}_{i+1} < r_\rho(\underline{v}_j) = \underline{v}_{j+1} < 1.$$

Since $v \in C_{\frac{p}{q}}$, $v_1 < v_{i+1} < v_{j+1} < 1$. By the definition of the Thurston map T, $v_1 < f_t(v'_i) < f_t(v'_j) < 1$. Note that the left branch of f_t, $(f_t)_- : [0, c(t)) \to [v_1, 1)$ is strictly increasing. Hence $v'_i < v'_j$.

II $c_\rho = \underline{v}_{q-1} \le \underline{v}_i < \underline{v}_j$. Similar to case I, this implies that

$$r_\rho(\underline{v}_i) = \underline{v}_{i+1} < r_\rho(\underline{v}_j) = \underline{v}_{j+1} < r_\rho(1) = \underline{v}_1.$$

Then $0 \le v_{i+1} < v_{j+1} < v_1$, which is equivalent to $0 \le f_t(v'_i) < f_t(v'_j) < v_1$. Since the branch $f_t : [c(t), 1) \to [0, v_1)$ is strictly increasing, $v'_i < v'_j$.

III $\underline{v}_i < c_\rho = \underline{v}_{q-1} \le \underline{v}_j$. Then $\underline{v}_1 < \underline{v}_{i+1} < 1, 0 \le \underline{v}_{j+1} < \underline{v}_1$, i.e.,

$$0 \le \underline{v}_{j+1} < \underline{v}_1 < \underline{v}_{i+1} < 1.$$

Hence $0 \le v_{j+1} < v_1 < v_{i+1} < 1$. $f_t(v'_j) \in [0, v_1)$ implies that v'_j is in the right branch, in particular $v'_j \ge c(t)$; similarly $f_t(v'_i) \in [v_1, 1)$ implies that $v'_i < c(t)$. Hence $v'_i < v'_j$.

□

Lemma 2.4.11 *The Thurston map $T : C_{\frac{p}{q}} \to C_{\frac{p}{q}}$ is continuous.*

Proof Fix $v = (0, v_1, \ldots, v_{q-2}, v_{q-1}) \in C_{\frac{p}{q}}$. Notice that

$$Tv = (0, v'_1, \ldots, v'_{q-1}) = (0, f_{v_1}^{-1}(v_2), \ldots, f_{v_1}^{-1}(v_{q-1}), c(v_1)).$$

Since both branches of f_{v_1} are homeomorphisms, $f_{v_1}^{-1}(x)$ is continuous as long as x is away from v_1. That is, for another configuration $\tilde{v}$, $T\tilde{v}$ stays close to Tv if the small perturbation from v to $\tilde{v}$ happens only in v_i for $i = 2, 3, \ldots, q-1$ (i.e. $\tilde{v}_1 = v_1$), and $|\tilde{v}_i - v_i| < \min_{2 \le i \le q-1}\{|v_1 - v_i|\}$ for all i.

Now let $\tilde{v} = (0, \tilde{v}_1, v_2, \ldots, v_{q-1})$ with $|\tilde{v}_1 - v_1| < \min_{2 \le i \le q-1}\{|v_1 - v_i|\}$. Then by the continuity of the family, $\tilde{v}'_i = f_{\tilde{v}_1}^{-1}(v_i)$ stays close to $v'_i = f_{v_1}^{-1}(v_i)$ for $i = 2, 3, \ldots, q-2$. For $i = q-1$, $\tilde{v}'_{q-1} = c(\tilde{v}_1)$ stays close to $v'_{q-1} = c(v_1)$ by the continuity of the function $c(t)$.

Hence we can prove that T is continuous. Let $\tilde{v}$ be a configuration close to v, more precisely let $\tilde{v} = (0, \tilde{v}_1, \tilde{v}_2, \ldots, \tilde{v}_{q-1})$ be such that $|\tilde{v}_i - v_i| < \min_{2 \le i \le q-1}\{|v_1 - v_i|\}$ for all $i = 1, 2, \ldots, q-1$. Then by triangle inequality we have

$$\begin{aligned}|T\tilde{v} - Tv| &= \left|T(0, \tilde{v}_1, \tilde{v}_2, \ldots, \tilde{v}_{q-1}) - T(0, v_1, v_2, \ldots, , v_{q-1})\right| \\ &\leq \left|T(0, \tilde{v}_1, \tilde{v}_2, \ldots, \tilde{v}_{q-1}) - T(0, v_1, \tilde{v}_2, \ldots, \tilde{v}_{q-1})\right| \\ &\quad + \left|T(0, v_1, \tilde{v}_2, \ldots, \tilde{v}_{q-1}) - T(0, v_1, v_2, \ldots, v_{q-1})\right|\end{aligned}$$

is small because of previous discussions. □

Proposition 2.4.12 *The Thurston map $T : C_{\frac{p}{q}} \to C_{\frac{p}{q}}$ has a fixed point.*

Proof Consider the orbit of zero under the rotation $r_{\frac{p}{q}} O_{\frac{p}{q}} = \left\{\underline{v}_i | i \in \{0, 1, \ldots, , q-1\}\right\}$. Let $0 < m \leq q-1$, such that $\underline{v}_m = \max\left\{\underline{v}_i\right\}$, the highest point. For $i > 0$ the lower neighbour of v_i has index l_i. In particular, let l_i such that $\underline{v}_{l_i} = \max\left\{\underline{v}_j | \underline{v}_j < \underline{v}_i\right\}$. For $i \neq m$ the upper neighbour of v_i has index u_i. So, let u_i be such that $\underline{v}_{u_i} = \min\left\{\underline{v}_j | \underline{v}_j > \underline{v}_i\right\}$. We describe now the boundary $\partial C_{\frac{p}{q}}$ of $C_{\frac{p}{q}}$. For this aim, consider the boundary components

$$B = \left\{v \in \overline{C}_{\frac{p}{q}} | v_m = 1\right\}.$$

For every $u \geq 1$ and l such that $l = l_u$, let

$$\partial_{l,u} = \left\{v \in \overline{C}_{\frac{p}{q}} | v_l = v_u\right\}.$$

Observe that

$$\partial C_{\frac{p}{q}} = B \cup \bigcup_{l,u} \partial_{l,u}$$

and because of the fact that for any $v \in B$ we have $v_0 = 0 \neq 1 = v_m$,

$$B \cap \bigcap_{l,u} \partial_{l,u} = \emptyset. \tag{2.4.12}$$

The next step is to extend the Thurston map to $\partial C_{\frac{p}{q}}$. Let $v \in \partial C_{\frac{p}{q}}$ and $t \in [0, 1)$ such that $f_t(0) = v_1 \mod 1$. We need to consider five different boundary components.

I $v \in \partial_{l,u}$ with $l, u \neq 1$ and $l \neq 0$. We define $Tv := v'$ as before, in particular $v'_{l-1} = f_t^{-1}(v_l) = f_t^{-1}(v_u) = v'_{u-1}$. As a consequence T extends to $\partial_{l,u}$ and

$$T : \partial_{l,u} \to \partial_{l-1,u-1}.$$

II $v \in \partial_{1,u_1}$. Observe that $r_{\frac{p}{q}}(\underline{v}_{u_0}) = \underline{v}_{u_1}$. We define $Tv := v'$ as before, in particular $v'_0 = 0 = f_t^{-1}(v_1) = f_t^{-1}(v_{u_1}) = v'_{u_0}$. As a consequence T extends to ∂_{1,u_1} and

$$T : \partial_{1,u_1} \to \partial_{0,u_0}.$$

III $v \in \partial_{l_1,1}$. Observe that $r_{\frac{p}{q}}(\underline{v}_m) = \underline{v}_{l_1}$. In this case $0 = f_t^{-1}(v_1) = f_t^{-1}(v_{l_1})$, we define $Tv := v'$ where $v'_0 = f_t^{-1}(v_1) = 0$, $v'_m = 1$ and other elements as before. As a consequence T extends to $\partial_{l_1,1}$ and

$$T : \partial_{l_1,1} \to B.$$

IV $v \in B$. Observe that $r_{\frac{p}{q}}(\underline{v}_{l_{q-1}}) = \underline{v}_m$. We define naturally $f_t^{-1}(1) := c(t)$ and then define $Tv := v'$ as before. In particular $v'_{l_{q-1}} = f_t^{-1}(v_m) = f_t^{-1}(1) = c(t) = v'_{q-1}$. As a consequence T extends to B and

$$T : B \to \partial_{l_{q-1},q-1}.$$

V $v \in \partial_{0,u_0}$. Observe that $r_{\frac{p}{q}}(\underline{v}_{u_{q-1}}) = \underline{v}_{u_0}$. We define $Tv := v'$ as before, in particular $v'_{q-1} = f_t^{-1}(v_0) = c(t) = f_t^{-1}(v_{u_0}) = v'_{u_{q-1}}$. As a consequence T extends to ∂_{0,u_0} and

$$T : \partial_{0,u_0} \to \partial_{q-1,u_{q-1}}.$$

The definition of the extension of T to $\partial C_{\frac{p}{q}}$ and Lemma 2.4.11 show that $T : \overline{C}_{\frac{p}{q}} \to \overline{C}_{\frac{p}{q}}$ is continuous.

By the definition of the extension of T to $\partial C_{\frac{p}{q}}$ we see that $\partial C_{\frac{p}{q}}$ is invariant under T, namely $T(\partial C_{\frac{p}{q}}) \subset \partial C_{\frac{p}{q}}$. In fact, by case IV we know that a point in B is mapped into component $\partial_{l_{q-1},q-1}$, and then we apply case I. Note that $l_{q-1} < q-1$, by case I the left index in $\partial_{l,u}$ will reach 1 before the right index. Note that for any pair l, u by definition we have $u = u_l$, the point will be iterated under T for $l_{q-1} - 1$ times in case I and end in component ∂_{1,u_1}, namely,

$$B \to \partial_{l_{q-1},q-1} \to \partial_{l_{q-1}-1,q-2} \to \cdots \to \partial_{1,u_1}.$$

Now applying the cases II and V we get a point in $\partial_{q-1,u_{q-1}}$. Using similar argument as before we get

$$\partial_{q-1,u_{q-1}} \to \partial_{q-2,u_{q-2}} \to \cdots \to \partial_{l_1,1}.$$

Then applying case III we see that the point ends up in B. In summary we have the following diagram:

$$\begin{aligned} B &\to \partial_{l_{q-1},q-1} \to \partial_{l_{q-1}-1,q-2} \to \cdots \\ &\to \partial_{1,u_1} \to \partial_{0,u_0} \to \partial_{q-1,u_{q-1}} \to \partial_{q-2,u_{q-2}} \to \cdots \\ &\to \partial_{l_1,1} \to B. \end{aligned} \tag{2.4.13}$$

Being $\overline{C}_{\frac{p}{q}}$ a ball we apply the Brouwer Fixed Point to the extension of T and we get a fixed point v^*. It remains to prove that $v^* \notin \partial C_{\frac{p}{q}}$. By contradiction, suppose that $v^* \in \partial C_{\frac{p}{q}}$. The diagram (2.4.13) together with (2.4.12) leads to a contradiction. Namely

$$v^* \in B \cap \bigcap_{l,u} \partial_{l,u} = \emptyset.$$

□

Remark 2.4.13 In the original holomorphic context Thurston showed that the Thurston map is a contraction and has a unique fixed point, see [MT88]. In the smooth context where the cutting point is also a critical point, i.e. $Df_t(c(t)) = 0$, the Thurston map near $\partial_{l_{q-1},q-1}$ is expanding. This property allows to construct an invariant compact ball in the configuration space to which the Brouwer Fixed Point Theorem can be applied. In our case we cannot rely on the expansivity because we consider families of homeomorphisms. The combinatorics of circle dynamics implies that the Thurston map has cyclic behaviour in the boundary as is shown in the diagram (2.4.13). This is the property for showing that the fixed configuration is not in the boundary. We expect that this technique can be applied in other non-smooth contexts.

Lemma 2.4.14 *If $v^* = (0, v_1^* = t^*, v_2^* \ldots, c(t^*))$ is a fixed point of $T : C_{\frac{p}{q}} \to C_{\frac{p}{q}}$, then the rotation number $\rho(f_{t^*}) = \frac{p}{q}$.*

Proof Recall that for all parameters $t \in [0, 1)$, $f_t(c(t)) = 0$. In particular $f_{t^*}(v_{q-1}^*) = 0$. Moreover being v^* a configuration, v_i^* for $i \in \{0, 1, \ldots, q-1\}$ are pairwise distinct.

Now, by $v_1^* = t^*$ we get $f_{t^*}(0) = v_1^*$ and being v^* a fixed point, $f_{t^*}(v_i^*) = v_{i+1}^*$ for $i = 1, 2, \ldots, q-2$. Hence, v^* is a periodic orbit of 0 under f_{t^*} of period q. For proving that $\rho(f_{t^*}) = \frac{p}{q}$ we apply Lemma 2.1.6 to the rotation $r_{\frac{p}{q}}$, substitute $n = q$ and $x = 0$, to see that

$$\#\left\{i \in \{0, 1, \ldots, q-1\} \,|\, \underline{v}_i \in [c_\rho, 1)\right\} = F^q(0) - r^q(0) = p - 0 = p.$$

Let

$$\varphi(x) = \begin{cases} 0 & x \in [0, c(t^*)) \\ 1 & x \in [c(t^*), 1) \end{cases}$$

then

$$\sum_{i=0}^{q-1} \varphi(f_{t^*}^i(0)) = \#\{i \in \{0, 1, \ldots, q-1\} | v_i^* \in [c(t^*), 1)\} = p,$$

where the last equality follows from v^* being in the configuration space and thus has the same combinatorial information as $\underline{O}_{\frac{p}{q}}$. Recall $f_{t^*}^q(0) = 0$. Hence, according to Lemma 2.1.6,

$$F^q(0) = p$$

and consequently $\rho(f_{t^*}) = \frac{p}{q}$. □

Proof of Proposition 2.4.9 The proof is obtained by combining Proposition 2.4.12 and Lemma 2.4.14. □

Proof of Theorem 2.4.1 Let $\rho \in [0, 1) \setminus \mathbb{Q}$ and let $(\rho_n)_{n\in\mathbb{N}} \in \mathbb{Q} \cap [0, 1)$ such that $\rho_n \to \rho$. By Proposition 2.4.9, there exists a sequence $(t_n)_{n\in\mathbb{N}} \in [0, 1)$ such that $\rho(f_{t_n}) = \rho_n$. Without loss of generality we may assume that $t_n \to t$ for some $t \in [0, 1)$ then by the continuity of the rotation number (see Proposition 2.4.3), $\rho(f_t) = \rho$. □

In this situation the Thurston method is overkill. The classical method is much shorter and simple. Our aim is just to present another method which is much more general and applicable to different situations.

2.5 Exercises

Exercise 2.5.1 Let X be a compact space, $f : X \to X$ a continuous dynamical system and $x \in X$. Prove that the limit set is invariant under f, i.e. $f(\omega(x)) = \omega(x)$.

Exercise 2.5.2 Let A be a minimal set of the dynamical system $f : X \to X$, f continuous and X compact. Prove that A is closed, invariant under f and that it does not have other proper subsets with the same properties.

Exercise 2.5.3 Prove Proposition 2.1.1. Hint. Consider $\tilde{f}$. In case $\tilde{f}$ is differentiable, let $F(x) = \tilde{f}(0) + \int_0^x D\tilde{f}$. Observe, every continuous function can be approximated by differentiable functions by using convolutions with narrow differentiable bumps.

Exercise 2.5.4 Prove Proposition 2.1.2.

Exercise 2.5.5 Prove Lemma 2.1.4

Exercise 2.5.6 Given an orientation preserving circle homeomorphism which has only periodic points. Show that f is a rotation.

Exercise 2.5.7 Let f be a decreasing circle homeomorphism. Prove that the limit set of any point is a fixed point or a periodic orbit of period two.

Exercise 2.5.8 Prove Proposition 2.2.5.

Exercise 2.5.9 Consider an orientation preserving circle homeomorphism f with rotation number $\rho(f) = \frac{p}{q}$. Show that for all $x \in \mathbb{S}^1$, $\omega(x)$ is a periodic orbit of period q. [Hint: study the dynamics of f^q].

Exercise 2.5.10 Prove Remark 2.3.2.

Exercise 2.5.11 Prove Proposition 2.3.4. [Hint: Consider the lifts F of f and G of g. Prove that the lift H of the semi-conjugacy h is a semi-conjugacy for F and G. In particular $HF^n = G^n H$].

Exercise 2.5.12 Show that, any semi-conjugacy between circle homeomorphisms with irrational rotation number is increasing.

Exercise 2.5.13 Consider two conjugate systems, f and g. Suppose that f has three periodic orbits of period 5. Show that g has three periodic orbits of period 5. Similarly, suppose that g has a minimal Cantor set. Show that f has a minimal Cantor set.

Exercise 2.5.14 Consider two conjugate systems, $f, g : [0, 1] \to [0, 1]$. Suppose that f has an attractor A, with respect to the Lebesgue measure, which is a Cantor set. Does g has a Cantor attractor?

Exercise 2.5.15 This exercise is an alternative discussion of rotation number and the classical Theorem by Poincaré (2.3.5). Let f be an orientation preserving circle homeomorphism without periodic orbits. Let M be the collection of invariant measures which occurs as weak limits of sequences

$$\mu_n = \frac{1}{n}\sum_{i=0}^{n-1} \delta_{f^i(x)}$$

with $x \in \mathbb{S}^1$. Show that M consists of a unique measure. The map f is said to be uniquely ergodic. As in the proof of Theorem 2.3.5 we get a semi-conjugacy with an uniquely defined irrational rotation. This can also be used to define the rotation number.

Hint: Study M when f is an irrational rotation. In particular show that for every $\mu \in M$ the map

$$h(x) = \int_0^x d\mu.$$

is differentiable with $Dh(x) = 1$.

Exercise 2.5.16 Given an orientation preserving circle homeomorphism f which has only periodic points. Show that if $f \prec \tilde{f}$ then $\rho(f) < \rho(\tilde{f})$.

Exercise 2.5.17 Prove Proposition 2.4.7.

Exercise 2.5.18 Sketch the configuration spaces $C_{\frac{3}{5}}$ and $C_{\frac{3}{4}}$.

Exercise 2.5.19 Apply the Thurston Method to the family (2.4.2) and show that there exists a map whose rotation number is $\frac{3}{4}$.

Exercise 2.5.20 Apply the Thurston Method to the family (2.4.2) and write the configuration space $C_{\frac{2}{7}}$ and the explicit formulas for the Thurston map.

Exercise 2.5.21 Apply the Thurston Method to the family (2.4.2) and write a little computer program to iterate the Thurston map $T : C_{\frac{2}{7}} \to C_{\frac{2}{7}}$. Observe that for every $v \in C_{\frac{2}{7}}$,

$$T^n v \to v^*.$$

Exercise 2.5.22 Is it possible to define a metric on the configuration space for which the Thurston map for the family (2.4.2) becomes a contraction?

Exercise 2.5.23 Given the family from (2.4.2) and $\rho = \frac{p}{q}$. Is the corresponding Thurston map a homeomorphism?

Exercise 2.5.24 Show that every irrational rotation number occurs only once in the family (2.4.2).

Chapter 3
Topology of Circle Diffeomorphisms

In the previous chapter we proved that any orientation preserving circle homeomorphism f having irrational rotation number ρ is semi-conjugate to the rotation of angle ρ, i.e. there exists a continuous surjective circle map h such that $h \circ f = r_\rho \circ h$. A natural question arises at this point: is such h an homeomorphism? In other words: is f conjugate to the rotation r_ρ? The answer is in general negative. To ensure the existence of the conjugacy we will need to impose some smoothness conditions on the function: instead of homeomorphisms we will consider circle diffeomorphisms. The motivation is that in the case of diffeomorphisms we can control how iterates distort the geometry.

The problem of the existence of a conjugacy is strictly connected with one topological question, namely the existence of wandering intervals.

3.1 Wandering Intervals

In this section f, as before, will denote an orientation preserving circle homeomorphism with irrational rotation number ρ. The only obstruction for the semi-conjugacy not to be a homeomorphism, i.e. a conjugacy, is the presence of wandering intervals.

Definition 3.1.1 We say that an interval $I \subset \mathbb{S}^1$ is a wandering interval if for all $n, m \in \mathbb{N}$, with $n \neq m$, $f^n(I) \cap f^m(I) = \emptyset$.

Lemma 3.1.2 *Let $h : \mathbb{S}^1 \to \mathbb{S}^1$ be a semi-conjugacy between f and r_ρ. The interval $I \subset \mathbb{S}^1$ is a wandering interval if and only if there exists $x \in \mathbb{S}^1$ such that $I \subset h^{-1}(x)$.*

Proof Suppose that there exists $x \in \mathbb{S}^1$ such that $I \subset h^{-1}(x)$. Then for all $n \in \mathbb{N}$, $h(f^n(I)) = r_\rho^n(x)$. If there exist $n, m \in \mathbb{N}$ such that $n \neq m$ and $f^n(I) \cap f^m(I) \neq \emptyset$,

Y. Dong et al., *One-Dimensional Dynamics*, Lecture Notes of the Unione Matematica Italiana 30, https://doi.org/10.1007/978-3-031-94882-4_3

then $r_\rho^n(x) = r_\rho^m(x)$. In particular x is periodic and by Lemma 2.2.6, $\rho \in \mathbb{Q}$. The contradiction proves that I is a wandering interval.

For the other implication we take a wandering interval I. Without loss of generality we may assume $I = (a, b)$ and we will prove that $h(I)$ is a point. By contradiction, suppose that $h(I) = (h(a), h(b))$ with $h(a) \neq h(b)$ (here we are using the fact that h is increasing, see Remark 2.3.6). Because $\rho \notin \mathbb{Q}$, by Proposition 2.2.7 there exists $n \in \mathbb{N}$, $n \neq 0$ such that

$$h(a) < r_\rho^n(h(a)) < h(b).$$

We use again that h is increasing and we get that

$$f^n(a) \in h^{-1}(r_\rho^n(h(a))) \subset I.$$

Since f is continuous $f^n(a) \in I$ implies that $f^n(I) \cap I \neq \emptyset$. This contradicts the fact that I is a wandering interval. □

The following corollary is a direct consequence of Lemma 3.1.2.

Corollary 3.1.3 *The following are equivalent:*

- *The map f has no wandering interval.*
- *The map f is conjugated to the rotation r_ρ.*
- *All orbits of f are dense.*

3.2 Distortion

From this section on, we will assume that our function f is an orientation preserving piecewise C^2 circle homeomorphism of the following form. We will consider the representation of f on the interval $[0, 1)$ introduced in (2.1.1). In this case the two diffeomorphic branches of f, $f_- = f|_{[0,c)}$ and $f_+ = f|_{[c,1)}$ extend to two twice continuously differentiable maps respectively in $[0, c]$ and $[c, 1]$. We will use $D\varphi(x)$ and $D^2\varphi(x)$ to denote the first and the second derivative respectively of a C^2 diffeomorphism φ at the point x.

Definition 3.2.1 Let $I \subset [0, 1)$ be an interval and let $\varphi : I \to \varphi(I)$ a C^2 diffeomorphism. The distortion of φ is defined as

$$Dist(\varphi) = \max_{x,y \in I} \log \frac{D\varphi(x)}{D\varphi(y)}.$$

The distortion measures how much the derivative of φ varies. Notice that if φ is an affine map then the distortion is always zero. We introduce now a function called "non-linearity" which will allow us to control the distortion under high iterates.

Definition 3.2.2 Let $I \subset [0, 1)$ be an interval and let $\varphi : I \to \varphi(I)$ a C^2 diffeomorphism. The non-linearity of φ is a function $\eta_\varphi : I \to \mathbb{R}$ such that for all $x \in I$,

$$\eta_\varphi(x) = \frac{D^2\varphi(x)}{D\varphi(x)} = D \log D\varphi(x).$$

The chain rule formula holds for the non-linearity in the following way (the proof is left as an exercise, see Exercise 3.5.2):

Lemma 3.2.3 *Let $I, J \subset [0, 1)$ be two intervals and let $\varphi : I \to J$ and $\psi : J \to \psi(J)$ be two C^2 diffeomorphisms. Then for all $x \in I$,*

$$\eta_{\psi\circ\varphi}(x) = \eta_\psi(\varphi(x))D\varphi(x) + \eta_\varphi(x).$$

Proposition 3.2.4 *Let $I \subset [0, 1)$ be an interval for which there exists $n \in \mathbb{N}$ such that $f^n : I \to f^n(I)$ is continuous, i.e., $f^i(I)$ doesn't contain the cutting point for all $i < n$. Suppose that, for $i < n$ the intervals $f^i(I)$ are pairwise disjoints. Then*

$$Dist(f^n|_I) \leq \int_{[0,1]} |\eta_f| dx.$$

Proof Notice that

$$Df^n(x) = \prod_{i=0}^{n-1} Df(f^i(x)).$$

As consequence

$$\begin{aligned}
Dist(f^n|_I) &= \max_{s,t\in I} \left|\log \frac{Df^n(s)}{Df^n(t)}\right| \\
&= \max_{s,t\in I} \left|\sum_{i=0}^{n-1} \log \frac{Df(f^i(s))}{Df(f^i(t))}\right| \\
&\leq \max_{s,t\in I} \sum_{i=0}^{n-1} \left|\int_{f^i(s)}^{f^i(t)} \eta_f\right| \\
&\leq \sum_{i=0}^{n-1} \int_{I_i=f^i(I)} |\eta_f| \\
&\leq \int |\eta_f|.
\end{aligned}$$

□

3.3 Conjugation of Circle Diffeomorphisms

In this section we will prove the following theorem by Denjoy.

Theorem 3.3.1 *Let f be an orientation preserving piecewise C^2 circle diffeomorphism with irrational rotation number ρ. Then f is conjugate to the rotation of angle ρ, r_ρ, i.e. there exists a circle homeomorphism h such that $h \circ f = r_\rho \circ h$.*

To prove Theorem 3.3.1 we need the following lemma:

Lemma 3.3.2 *Let I be a wandering interval, then there exists a sequence of closest approaches, i.e. a sequence of positive time $t_n \to \infty$ such that $f^i(I) \cap [f^{t_n}(I), I] = \emptyset$[1] for all $-t_n < i < t_n, i \neq 0$.*

Proof Since I is a wandering interval, by Lemma 3.1.2 there exists $x \in S^1$ such that $I \subset h^{-1}(x)$. To simplify the notation, for all $i \in \mathbb{Z}$ we use the notation $x_i := f^i(x)$ and $d_i := |x_i - x|$. For all $n \in \mathbb{N}$ let $K_n := \{j \in \mathbb{Z} | -n < j < n, j \neq 0\}$.

We first prove that there exists a sequence $t_n \to \infty$ of closest approaches to x for the rotation r_ρ. Namely for $x \in S^1$, we construct a sequence of positive time $t_n \to \infty$ such that $d_i > d_{t_n}$ for all $i \in K_{t_n}$. In particular, $x_i \notin [x - d_{t_n}, x + d_{t_n}]$ for all $i \in K_{t_n}$. The proof is done by induction.

Notice that $t_1 := 1$ satisfies the condition trivially. Suppose we have $t_n > 0$ satisfying $d_i > d_{t_n}$ for all $i \in K_{t_n}$. By Proposition 2.2.7, the orbit $\{x_i\}_{i \geq 0}$ under irrational rotation r_ρ is dense, hence there exist points of the orbit in $[x - d_{t_n}, x + d_{t_n}]$. We define t_{n+1} as the first time when the orbit lands in the interval, i.e., $d_i > d_{t_n}$ for all $0 < i < t_{n+1}$ and $d_{t_{n+1}} < d_{t_n}$, then we have $d_i > d_{t_{n+1}}$ for all $0 < i < t_{n+1}$. We want to show that it also holds for all $-t_{n+1} < i < 0$. Assume by contradiction that $d_k < d_{t_{n+1}}$ for some $-t_{n+1} < k < 0$, then since the Lebesgue measure is invariant under rotation, we have

$$d_{-k} = |x_{-k} - x| = |x - x_k| = d_k < d_{t_{n+1}},$$

which contradicts with the definition of t_{n+1} as $0 < -k < t_{n+1}$. Hence we have proved that $d_i > d_{t_{n+1}}$ for all $i \in K_{t_{n+1}}$.

Now we prove that a time t_n in the above-constructed sequence is also a time of closest approach to I for f, i.e., $f^i(I) \cap [f^{t_n}(I), I] = \emptyset$ for all $i \in K_{t_n}$. Without loss of generality we may assume x_{t_n} is to the right of x, then we have $x_i \notin [x, x_{t_n}]$ for all $i \in K_{t_n}$. Using the definition of semi-conjugacy and by our choice of x and I we have

$$h^{-1}(x_i) = h^{-1}(r_\rho^i(x)) = h^{-1}(r_\rho^i(h(I))) = h^{-1}(h(f^i(I))) \supset f^i(I).$$

[1] The notation $[f^{t_n}(I), I]$ represents the smallest interval containing $f^{t_n}(I)$ and I.

Notice that by Remark 2.3.6, the semi-conjugacy h is increasing. Hence h^{-1} is increasing, namely

$$x > y \Rightarrow \inf\{h^{-1}(x)\} > \sup\{h^{-1}(y)\}, \quad \forall x, y \in \mathbb{S}^1.$$

Since we assume $x < x_{t_n}$, the above statement implies that the corresponding interval $h^{-1}(x) \supset I$ is to the left of $h^{-1}(x_{t_n}) \supset f^{t_n}(I)$ and they are disjoint. Since $x_i \notin [x, x_{t_n}]$, if $x_i < x$ we have that the interval $f^i(I)$ is to the left of I and they are disjoint. In particular $f^i(I) \cap [f^{t_n}(I), I] = \emptyset$. Similarly we can prove that $f^i(I) \cap [f^{t_n}(I), I] = \emptyset$ also holds if $x_i > x_{t_n}$. □

Proof of Theorem 3.3.1 By Corollary 3.1.3, it is enough to prove that f does not have wandering intervals. By contradiction we suppose that there exists a maximal wandering interval I. By Lemma 3.1.2 and by the maximality of I, there exists $x \in \mathbb{S}^1$ such that $I = h^{-1}(x)$.

By Lemma 3.3.2, let $t_n \to \infty$ be a time of closest approach. For $n \in \mathbb{Z}$, let $I_n = f^n(I)$. The function

$$f^{t_n} : T_n = [I, I_{-t_n}] \to [I_{t_n}, I] \tag{3.3.1}$$

is an homeomorphism. We prove that the intervals $\{f^i(T_n) | 0 \leq i < t_n\}$ are pairwise disjoint. Suppose not, then there exists $0 < i < t_n$ such that $T_n \cap f^i(T_n) \neq \emptyset$. That is,

$$[I, I_{-t_n}] \cap [I_i, I_{-(t_n-i)}] \neq \emptyset.$$

This can only happen if at least one of the two ends of T_n intersects $f^i(T_n)$, or $f^i(T_n)$ is contained in T_n. In the first case, either $I_{-t_n} \cap [I_i, I_{-(t_n-i)}] \neq \emptyset$ or $I \cap [I_i, I_{-(t_n-i)}] \neq \emptyset$. Applying f^{t_n-i} we get $I_{-i} \cap [I_{t_n}, I] \neq \emptyset$ or $I_{t_n-i} \cap [I_{t_n}, I] \neq \emptyset$, both of which lead to contradiction with the definition of t_n. In the second case $I_{-(t_n-i)} \subset T_n = [I, I_{-t_n}]$, applying f^{t_n} we get $I_i \cap [I_{t_n}, I] \neq \emptyset$ which also contradicts with the fact that t_n is a closest approach.

The pairwise disjointness implies that there exists a constant C such that for all $n \in \mathbb{N}$ $Dist(f^{t_n}|_{T_n}) \leq C$, see Proposition 3.2.4. The intervals $\{I_n | n \in \mathbb{Z}\}$ are pairwise disjoint, then $|I_{t_n}|, |I_{-t_n}| \to 0$. As a consequence,

$$C \geq Dist(f^{t_n}|_{T_n}) \geq \left| \log \frac{\frac{|f^{t_n}(I)|}{|I|}}{\frac{|f^{t_n}(I_{-t_n})|}{|I_{-t_n}|}} \right| = \left| \log \frac{|I_{t_n}|}{|I|} \frac{|I_{-t_n}|}{|I|} \right| \to \infty.$$

The contradiction comes from the fact that f^{t_n} contracts strongly on I and expands strongly on I_{-t_n} which is not allowed by the boundedness of the distortion. □

3.4 Denjoy Counterexample

In Theorem 3.3.1 we proved that any C^2 orientation preserving circle diffeomorphism having irrational rotation number does not have wandering intervals and it is therefore conjugate to an irrational rotation. In this section we prove that the hypothesis on the smoothness of the function is essential, namely for all irrational numbers ρ we construct a C^1 orientation preserving circle diffeomorphism having rotation number ρ which has a wandering interval. Here is the precise statement of the theorem.

Theorem 3.4.1 *Let $\rho \in [0, 1)$ an irrational number, then there exists a C^1 orientation preserving circle diffeomorphism f with the following properties:*

- $\rho(f) = \rho$,
- *f has a wandering interval,*
- *f is not conjugated to a rotation.*

Proof Fix $\rho \in [0, 1)$ an irrational number and consider the representation of the rotation of angle ρ in the interval $[0, 1)$, $r_\rho : [0, 1) \to [0, 1)$ such that $r_\rho(x) = x+\rho \mod 1$. Let

$$O_\rho = \left\{x_i = r_\rho^i(0) | i \in \mathbb{Z}\right\}$$

the orbit of 0 under the rotation of angle ρ and define a probability measure

$$\mu = \frac{1}{M}\left[\lambda + \sum_{i\in\mathbb{Z}} \frac{1}{i^2+1}\delta_{x_i}\right]$$

where $M = 1 + \sum_{i\in\mathbb{Z}} \frac{1}{i^2+1}$, λ is the Lebesgue measure and δ_x is the point mass at x.

We define a function $h_{-1} : [0, 1) \to [0, 1)$ such that

$$h_{-1}(x) = \int_0^x d\mu.$$

Notice that h_{-1} is increasing and discontinuous at all x_i. We consider then $h : [0, 1) \to [0, 1)$, in essence the inverse of h_{-1}, which is a continuous increasing function such that for all $x \in [0, 1)$,

$$(h \circ h_{-1})(x) = x. \tag{3.4.1}$$

Observe that h has the following properties:

1. h is not strictly increasing, for all $i \in \mathbb{Z}$ there exist an interval I_i such that

$$I_i = h^{-1}(x_i).$$

Moreover,

$$|I_i| = \frac{1}{M} \cdot \frac{1}{i^2+1}.$$

2. for all measurable sets $A \in [0, 1)$,

$$|h^{-1}(A)| = \mu(A). \tag{3.4.2}$$

We define

$$S = [0, 1) \setminus \bigcup_{i \in \mathbb{Z}} I_i \tag{3.4.3}$$

and we observe that $h : S \to [0, 1) \setminus O_\rho$ is an homeomorphism. Our $\mathcal{C}^1$ orientation preserving circle diffeomorphism f is defined in the following way. For every $i \in \mathbb{Z}$ choose $f_i : I_i \to I_{i+1}$ a $\mathcal{C}^1$ orientation preserving diffeomorphism such that

1. $Df_i|_{\partial I_i} = 1$,
2. $|Df_i(x) - 1| \leq \frac{C}{|i|+1}$ for some constant C.

and we define $f : [0, 1) \to [0, 1)$ as

$$f(x) = \begin{cases} h_{-1} \circ r_\rho \circ h(x) & x \in S \\ f_i(x) & x \in I_i \end{cases}$$

Observe that the map f is an orientation preserving circle homeomorphism. We claim that f is indeed a $\mathcal{C}^1$ circle diffeomorphism and h is a semi-conjugacy between f and r_ρ.

(i) f is differentiable.

Suppose $x \in \bigcup_{i \in \mathbb{Z}} \overset{\circ}{I}_i$, then f is differentiable at x by construction.

Suppose $x \in S$. Fix $\varepsilon > 0$ and we will show that for $\delta > 0$ sufficiently small,

$$1 - \varepsilon \leq \frac{|f([x, x+\delta])|}{|[x, x+\delta]|} \leq 1 + \varepsilon. \tag{3.4.4}$$

Let δ_0 small enough such that for all $0 < \delta < \delta_0$, all integers k such that $I_k \cap [x, x+\delta] \neq \emptyset$ satisfy

$$1 - \varepsilon \leq \frac{k^2+1}{(k+1)^2+1} \leq 1 + \varepsilon$$

and

$$\frac{C}{|k|+1} \leq \varepsilon.$$

For any positive $\delta < \delta_0$, we denote $S^\delta := [x, x+\delta] \cap S$, $I^\delta := [x, x+\delta] \backslash S^\delta$, then we have

$$[x, x+\delta] = S^\delta \sqcup I^\delta.$$

To estimate $|f(S^\delta)|$, note that since h is a bijection on $S \supset f(S^\delta)$, we have

$$|f(S^\delta)| = |h^{-1}(h(f(S^\delta)))|.$$

Now using (3.4.2) and since $\mu|_{h(S)}$ is the Lebesgue measure λ which is invariant under the rotation r_ρ, we have

$$|f(S^\delta)| = \mu(h(f(S^\delta))) = \mu(h(h_{-1} \circ r_\rho \circ h(S^\delta))) = \mu(h(S^\delta)) = |S^\delta|. \quad (3.4.5)$$

To estimate $|f(I^\delta)|$, we denote $K_\delta := \{k \in \mathbb{Z} | I_k \subset [x, x+\delta]\}$ and let $J := I_{k_0} \cap [x, x+\delta]$ if $x+\delta \in I_{k_0}$ for some $k_0 \notin K_\delta$, otherwise let J be the empty set. Then we have

$$I^\delta = (\bigsqcup_{k \in K_\delta} I_k) \sqcup J.$$

Note that $J \subset I_{k_0}$, by the mean value theorem $|f(J)| = Df_{k_0}(\theta) \cdot |J|$ for some $\theta \in I_{k_0}$, together with our choice of δ we have

$$(1-\varepsilon)|J| \leq |f(J)| \leq (1+\varepsilon)|J|.$$

Hence

$$\begin{aligned}
|f(I^\delta)| &= \sum_{k \in K_\delta} |f(I_k)| + |f(J)| = \sum_{k \in K_\delta} |I_{k+1}| + |f(J)| \\
&= \frac{1}{M} \sum_{k \in K_\delta} \frac{1}{(k+1)^2+1} + |f(J)| = \sum_{k \in K_\delta} |I_k| \cdot \frac{k^2+1}{(k+1)^2+1} + |f(J)| \\
&\leq (1+\varepsilon) \sum_{k \in K_\delta} |I_k| + (1+\varepsilon)|J| = (1+\varepsilon)|I^\delta|,
\end{aligned} \quad (3.4.6)$$

and similarly

$$|f(I^\delta)| \geq (1-\varepsilon)|I^\delta|. \quad (3.4.7)$$

Using (3.4.5) and (3.4.6) we have

$$\frac{|f([x, x+\delta])|}{|[x, x+\delta]|} = \frac{|f(S^\delta \sqcup I^\delta)|}{|S^\delta| + |I^\delta|} = \frac{|f(S^\delta)| + |f(I^\delta)|}{|S^\delta| + |I^\delta|} \leq \frac{|S^\delta| + (1+\varepsilon)|I^\delta|}{|S^\delta| + |I^\delta|}$$

$$= 1 + \varepsilon \frac{|I^\delta|}{|S^\delta| + |I^\delta|} \leq 1 + \varepsilon.$$

Similarly using (3.4.5) and (3.4.7) we have

$$\frac{|f([x, x+\delta])|}{|[x, x+\delta]|} \geq 1 - \varepsilon.$$

Since $\varepsilon > 0$ is arbitrary, we have proved that the right derivative of f at points in S,

$$D_+ f(x) := \lim_{t \to x^+} \frac{|f(t) - f(x)|}{|t - x|} = 1, \quad \forall x \in S.$$

Similarly we can show that the left derivative $D_- f(x)$ at points $x \in S$ is also 1, hence f is differentiable at all point in S, and the derivative

$$Df(x) = 1, \quad \forall x \in S.$$

Suppose $x \in \bigcup_{i \in \mathbb{Z}} \partial I_i$. Without loss of generality we may assume x is the right end point of I_j for some $j \in \mathbb{Z}$. Then by construction of f we have the left derivative

$$D_- f(x) = Df_j|_{\partial I_j} = 1.$$

Using the same argument as for $x \in S$ we can show that the right derivative $D_+ f(x) = 1$, hence f is differentiable at all points in $\bigcup_{i \in \mathbb{Z}} \partial I_i$, and the derivative

$$Df(x) = 1, \quad \forall x \in \bigcup_{i \in \mathbb{Z}} \partial I_i.$$

We have proved that f is differentiable in $[0, 1)$.

(ii) f is continuously differentiable. By construction f is continuously differentiable on $\cup \mathring{I}_i$. Let $x \in S$ and $\{y_n\}_{n \in \mathbb{N}}$ a sequence converging to x, $y_n \to x$. Observe that if $y_n \in S$ then $Df(y_n) = 1$. If $y_n \in I_{i_n}$ then $|i_n| \to \infty$ and $|Df(y_n) - 1| = \mathcal{O}\left(\frac{1}{|i|+1}\right) \to 0$. Hence

$$Df(y_n) \to 1 = Df(x).$$

Similarly we prove that f is continuously differentiable in x with $x \in \partial I_i$.

(iii) We prove now that h is a semi-conjugacy between f and r_ρ, i.e.

$$h \circ f = r_\rho \circ h.$$

In fact,

- if $x \in S$, by (3.4.1), $h \circ f(x) = h \circ h_{-1} \circ r_\rho \circ h(x) = r_\rho \circ h(x)$,
- if $x \in I_i$ for some $i \in \mathbb{Z}$, then $f(x) \in I_{i+1}$; consequently $h \circ f(x) \in h(I_{i+1}) = \{x_{i+1}\}$, hence $h \circ f(x) = x_{i+1} = r_\rho(x_i) = r_\rho(h(I_i)) = r_\rho \circ h(x)$.

Now since h is the semi-conjugacation between f and r_ρ, by Lemma 3.1.2 I_0 is a wandering interval for f and by Proposition 2.3.4, $\rho(f) = \rho$. Using Corollary 3.1.3 we see that f is not conjugated to the rotation r_ρ. □

Remark 3.4.2 We would like to stress that, as shows Theorem 3.4.1, it is possible to find in $\mathcal{C}^1$ systems special behaviours which do not occur in systems of higher smoothness. In this course we are interested in the study of systems which are at least $\mathcal{C}^2$ regular. The reason is that models which represent natural processes are often written in the form of differential equations which leads to dynamical systems of higher smoothness.

3.5 Exercises

Exercise 3.5.1 *Prove that any semi-conjugacy between circle homeomorphisms with irrational rotation number, is increasing.*[2]

Exercise 3.5.2 *Prove Lemma 3.2.3.*

Exercise 3.5.3 *For every irrational $\rho \in [0, 1)$ one can construct a $\mathcal{C}^1$ circle diffeomorphism with rotation number ρ which has two different wandering intervals I_1 and I_2 such that $f^n(I_1) \cap f^m(I_2) = \emptyset$ for all n, m. There are two different wandering orbits. Indicate how the proof of the Denjoy Theorem can be adapted.*

Exercise 3.5.4 *Construct a $\mathcal{C}^1$ Denjoy-Counter example where the invariant Cantor set S, see (3.4.3), has positive Lebesgue measure.*

Exercise 3.5.5 *Let $f : [0, 1] \to [0, 1]$ be a $\mathcal{C}^2$ circle diffeomorphism for which there exists $n \in \mathbb{N}$ such that $f^n : I \to f^n(I)$ is an homeomorphism. For all $i < n$ denote by $I_i = f^i(I)$ and by*

$$d = \max_{x \in [0,1]} \#\{i < n | x \in I_i\}$$

the so called intersection multiplicity. Prove that

[2] This is a difficult exercise.

$$Dist(f_{|I}^{n}) \leq d \int_{[0,1]} |\eta_f| dx.$$

Exercise 3.5.6 *Let $f : \mathbb{S}^1 \to \mathbb{S}^1$ be a continuous order preserving circle map for which there exists an open arc U such that the following properties are satisfied:*

- $f(U) = c$,
- $f_{|\mathbb{S}^1 \setminus \overline{U}}$ *is a C^2 diffeomorphism in its image,*
- $\eta_f = \frac{D^2 f}{Df}$ *extends continuously to the boundary points of U.*

Prove that:

1. *the rotation number $\rho(f)$ of f exists and it is well defined,*
2. *show that f is semi-conjugated to a rotation over angle $\rho(f)$,*
3. *the images $\{f^i(U)\}$ are pairwise disjoint,*
4. *$C = \mathbb{S}^1 \setminus \bigcup_{i \geq 0} f^{i-}(U)$ does not contain intervals (this implies that C is a Cantor set and that f does not have wandering intervals different to U),*
5. *C has zero Lebesgue measure, $\lambda(C) = 0$.*

This is a more involved exercise, a little project. However the solution can be found adapting the ideas that we presented in the chapters so far.

Chapter 4
Renormalization

In the previous chapters we described the topological properties of circle diffeomorphisms, i.e. the order of the orbits, properties of the rotation number, non-existence of wandering intervals, density of the orbits, etc. We are now interested in studying the systems from a geometrical point of view. We learned that the orbits of diffeomorphisms are deformations of orbits of a rotation. Of course, on large scale we can deform the orbits in any possible way creating any kind of large scale geometry. For example, one can use a smooth conjugation to change the large scale geometrical properties. The natural question is then to ask whether it is possible to deform the geometry on small scales. A smooth conjugacy can not achieve that. For studying this question we need a microscope which allows us to look at the details of the dynamics at small scale. The mathematical object which will play the role of the microscope is the renormalization operator. In the following sections we introduce renormalization and discuss its properties. The geometrical use of renormalization will be discussed in the next chapter.

4.1 Definition of Renormalization

The renormalization operator will assign to a circle homeomorphism another circle homeomorphism. As we will see later, the new homeomorphism describes the dynamics of the original homeomorphism on a smaller scale. Unfortunately, when one applies the renormalization operator to a circle diffeomorphism, the resulting map might not be a diffeomorphism. For this reason we need to enlarge the space of circle diffeomoprhims. It will contain the so-called generaized circle diffeomorphisms. Indeed, renormalization will assign to each generalized circle diffeomorphism a generalized circle diffeomorphism.

A generalized circle diffeomorphisms is a piecewise circle diffeomorphisms whose two branches are $k \geq 2$ times differentiable with continuous derivative with

Y. Dong et al., *One-Dimensional Dynamics*, Lecture Notes of the Unione Matematica Italiana 30, https://doi.org/10.1007/978-3-031-94882-4_4

weakened connecting conditions at the endpoints, compare to the representaion of circle homeomorphisms given in (2.1.1). We consider the space $\mathcal{D}^k$ consisting of functions f of the following form:

- $f : [0, 1) \to [0, 1)$,
- there exist $c_f, v_f \in (0, 1)$ and two orientation preserving C^k diffeomorphisms $f_- : [0, c_f] \to [v_f, 1]$ and $f_+ : [c_f, 1] \to [0, v_f]$,
- $\int \eta_f = 0$, where $\eta_f(x) = \begin{cases} \eta_{f_-}(x), & x \in [0, c_f) \\ \eta_{f_+}(x), & x \in [c_f, 1) \end{cases}$,
- $f(x) = \begin{cases} f_-(x) \; x \in [0, c_f) \\ f_+(x) \; x \in [c_f, 1) \end{cases}$

The point c_f is called the cutting point of f. Note that the third condition allows that the map does not represent a circle diffeomorphism, i.e. the corresponding circle homeomorphism might have two points where it is not differentiable.

In the case that the map represents a circle diffeomorphism one obtains:

$$\begin{aligned} \int \eta_f &= \int_0^{c_f} \eta_{f_-} + \int_{c_f}^1 \eta_{f_+} \\ &= \log Df_-(c_f) - \log Df_-(0) + \log Df_+(1) - \log Df_+(c_f) \\ &= \log \frac{Df_+(1)}{Df_-(0)} + \log \frac{Df_-(c_f)}{Df_+(c_f)} = 0. \end{aligned}$$

It will turn out that the third condition is preserved under renormalization.

We define a distance on $\mathcal{D}^k$ in the following way. If $I \subset [0, 1)$ is an interval, then $A_I : [0, 1) \to I$ is the orientation preserving affine map. Let $I, J \subset [0, 1)$ be two intervals and let $\varphi : I \to J$ be a surjective map. The normalized version of φ, for which we use the notation $N\varphi : [0, 1) \to [0, 1)$, is defined as

$$N\varphi = A_J^{-1} \circ \varphi \circ A_I. \tag{4.1.1}$$

Let us now define a distance on $\mathcal{D}^k$. Let $f, g \in \mathcal{D}^k$, the distance, $0 \leq r \leq k$, will be

$$d_r(f, g) = \max\left\{|c_f - c_g|, |v_f - v_g|, dist_{C^r}(Nf_-, Ng_-), dist_{C^r}(Nf_+, Ng_+))\right\}$$

where $dist_{C^r}$ is the C^r distance, i.e. for two C^r interval diffeomorphisms φ, ψ,

$$dist_{C^r}(\varphi, \psi) = \max_{i=0,1,\ldots,r} \{ \sup_{x\in[0,1)} |D^i\varphi(x) - D^i\psi(x)|\}.$$

The d_r distance is closely related to the distance d between circle homeomorphisms defined in Remark 2.4.2. The proof of the following remark is left as Exercise 4.7.2.

Remark 4.1.1 The distance d defined in Remark 2.4.2 is dominated by the distance d_r, $0 \leq r \leq k$. More precisely, for all $\varepsilon > 0$ there exists $\delta > 0$ such that, for any two generalized circle diffeomorphisms $f, g \in \mathcal{D}^k$ satisfying $d_r(f, g) < \delta$, we have $d(f, g) < \varepsilon$.

Observe that the natural distance on $\mathcal{D}^k$ is d_k.

We consider now the three subsets of $\mathcal{D}^k$,

$$\mathcal{D}_-^k = \left\{ f \in \mathcal{D}^k | c_f < v_f \right\},$$

$$\mathcal{D}_+^k = \left\{ f \in \mathcal{D}^k | c_f > v_f \right\}$$

and

$$\mathcal{D}_0^k = \left\{ f \in \mathcal{D}^k | c_f = v_f \right\}.$$

We are ready to define renormalization

$$R : \mathcal{D}_-^k \cup \mathcal{D}_+^k \to \mathcal{D}^k.$$

For $f \in \mathcal{D}_-^k \cup \mathcal{D}_+^k$, we define

$$U_f = \begin{cases} [0, v_f) & f \in \mathcal{D}_-^k \\ [v_f, 1) & f \in \mathcal{D}_+^k \end{cases}.$$

For $x \in U_f$ let $k_x = \min \left\{ k > 0 | f^k(x) \in U_f \right\}$. Observe, k_x is the first time that the orbit of x returns to U_f. The pre-renormalization $pRf : U_f \to U_f$ is defined as

$$pRf(x) = f^{k_x}(x)$$

for $x \in U_f$. It is also called the first return map to U_f.

We have the following lemma about pre-renormalization (Exercise 4.7.3). The lemma is illustrated in the Figs. 4.1 and 4.2.

Lemma 4.1.2 *For all $f \in \mathcal{D}_-^k$ and $x \in U_f$,*

$$pRf(x) = \begin{cases} f^2(x) = f_+ \circ f_-(x) & x \in [0, c_f) \\ f(x) = f_+(x) & x \in [c_f, v_f) \end{cases}$$

For all $f \in \mathcal{D}_+^k$ and $x \in U_f$,

$$pRf(x) = \begin{cases} f(x) = f_-(x) & x \in [v_f, c_f) \\ f^2(x) = f_- \circ f_+(x) & x \in [c_f, 1) \end{cases}$$

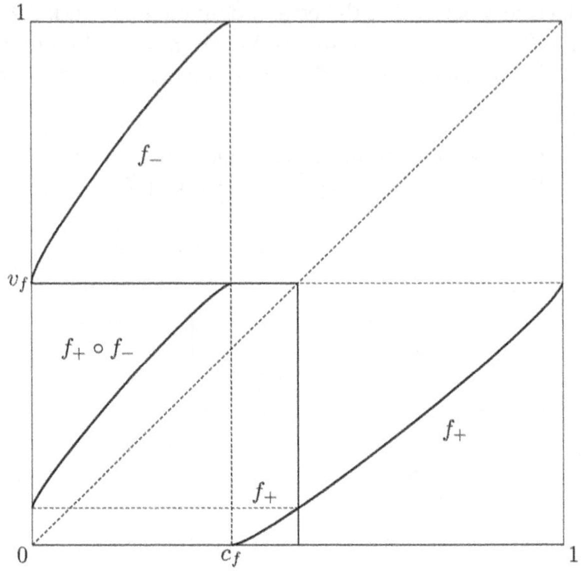

Fig. 4.1 The pre-renormalization of $f \in \mathcal{D}^k_-$

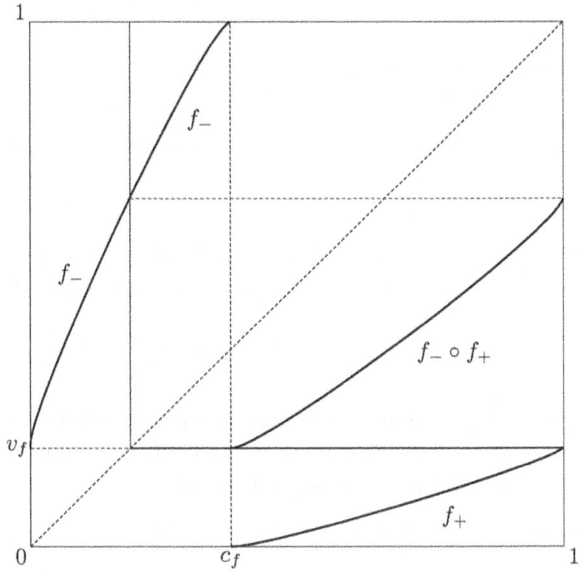

Fig. 4.2 The pre-renormalization of $f \in \mathcal{D}^k_+$

Definition 4.1.3 A function $f \in \mathcal{D}^k$ is renormalizable if $f \in \mathcal{D}^k_- \cup \mathcal{D}^k_+$. The renormalization is defined as

$$Rf = NpRf,$$

where $R : \mathcal{D}^k_- \cup \mathcal{D}^k_+ \to \mathcal{D}^k$ is called the renormalization operator.

The renormalization operator is well-defined (Exercise 4.7.4).

Definition 4.1.4 Let $n \in \mathbb{N}$ and $f \in \mathcal{D}^k$. We say that f is n-renormalizable if for all $i \in \mathbb{N}$, $i < n$, $R^i f$ is renormalizable, i.e. $R^i f \in \mathcal{D}^k_- \cup \mathcal{D}^k_+$. We say that f is ∞-renormalizable if it does not have periodic orbits. The space of ∞-renormalizable maps will be denoted by $\mathcal{I}^k$.

The proof of the following two lemmas are left as an exercise (Exercises 4.7.6 and 4.7.7):

Lemma 4.1.5 *If $f \in \mathcal{D}^k$ is m-renormalizable, then for all $n \leq m$ $R^n f$ is the first return map to an interval U_n up to scaling, where*

$$U_n = A_{U_f} \circ A_{U_{Rf}} \circ \cdots \circ A_{U_{R^{n-1}f}}([0, 1]).$$

Lemma 4.1.6 *Let $f \in \mathcal{D}^k$ be m-renormalizable. For all $n \leq m$, let $L_n = [0, c_f) \cap U_n$ be the left connected components of $U_n \setminus \{c_f\}$ and $R_n = [c_f, 1) \cap U_n$ the right one. Then there exist two positive numbers $l_n, r_n \in \mathbb{N}$ such that the first return function $F_n : U_n \to U_n$ to U_n has the following form:*

$$F_n(x) = \begin{cases} f^{l_n}(x) & x \in L_n \\ f^{r_n}(x) & x \in R_n \end{cases}$$

Observe that renormalization consists of two step: the first is the construction of the pre-renormalization, the first return map, and the second is the normalization. The first step describe the dynamics, it describes the first return map to a certain domain. It is the essential part. The role of the second step is to transform such a domain to the original one in order to make the renormalization again a map on the same domain as the original map. In a certain sense, this step is not as essential as the first. Nevertheless, the second step makes renormalization a dynamical systems on the space of generalized circle diffeomorphisms.

4.2 Dynamical Partitions

For every $n \in \mathbb{N}$, we consider the collection of the iterates of L_n and R_n, namely

$$\mathscr{L}_n = \left\{L_n, f(L_n), \ldots, f^{l_n-1}(L_n)\right\}$$

and

$$\mathscr{R}_n = \left\{R_n, f(R_n), \ldots, f^{r_n-1}(R_n)\right\}.$$

Their union,

$$\mathscr{P}_n = \mathscr{L}_n \cup \mathscr{R}_n$$

is called the dynamical partition of level n. The elements of $\mathscr{P}_n$ are called pieces.

Lemma 4.2.1 *Let $n \in \mathbb{N}$ and $f \in \mathcal{D}^k$ $(n+1)$-renormalizable. If $R^n f \in \mathcal{D}^k_-$, then:*

1. $l_{n+1} = l_n + r_n$,
2. $r_{n+1} = r_n$,
3. $L_{n+1} = L_n$,
4. $R_{n+1} = R_n \setminus f^{l_n}(L_n)$,
5. *for* $0 \le i < r_n$, $f^i(R_{n+1}) \cap f^i(f^{l_n}(L_n)) = \emptyset$,
6. *for* $0 \le i < r_n$, $f^i(R_n) = f^i(R_{n+1}) \cup f^i(f^{l_n}(L_n))$.

Proof The proof follows immediately from the definition, Lemmas 4.1.2 and 4.1.6. □

Lemma 4.2.2 *Let $n \in \mathbb{N}$ and $f \in \mathcal{D}^k$ $(n+1)$-renormalizable. If $R^n f \in \mathcal{D}^k_+$, then:*

1. $l_{n+1} = l_n$,
2. $r_{n+1} = r_n + l_n$,
3. $L_{n+1} = L_n \setminus f^{r_n}(R_n)$,
4. $R_{n+1} = R_n$,
5. *for* $0 \le i < l_n$, $f^i(L_{n+1}) \cap f^i(f^{r_n}(R_n)) = \emptyset$,
6. *for* $0 \le i < l_n$, $f^i(L_n) = f^i(L_{n+1}) \cup f^i(f^{r_n}(R_n))$.

Proof The proof follows immediately from the definition, Lemmas 4.1.2 and 4.1.6. □

Lemma 4.2.3 *The dynamical partition of level n, $\mathscr{P}_n$ has the following properties:*

1. *the pieces of $\mathscr{P}_n$ are pairwise disjoint,*
2. $\bigcup_{I \in \mathscr{P}_n} I = [0, 1]$.

Proof The proof is by induction. For $n = 0$, the Lemma is trivially true. Supposing that it is true for n, the proof for $n + 1$ follows immediately from Lemmas 4.2.1 and 4.2.2. □

Remark 4.2.4 Observe that $\mathscr{P}_{n+1}$ is a refinement of $\mathscr{P}_n$. In particular, if $R^n f \in \mathcal{D}^k_-$, then

$$\mathscr{R}_{n+1} = \left\{ f^i(R_{n+1}) \mid 0 \le i < r_n \right\}$$

and

$$\mathscr{L}_{n+1} = \mathscr{L}_n \cup \left\{ f^i(f^{l_n}(L_n)) \mid 0 \le i < r_n \right\}.$$

If $R^n f \in \mathcal{D}^k_+$, then

$$\mathcal{R}_{n+1} = \mathcal{R}_n \cup \left\{ f^i(f^{r_n}(R_n)) \mid 0 \le i < l_n \right\}$$

and

$$\mathcal{L}_{n+1} = \left\{ f^i(L_{n+1}) \mid 0 \le i < l_n \right\}.$$

The partition $\mathcal{P}_n$ is rather complicated. Schematically one can represent the dynamical partitions as *towers*, see Figs. 4.3 and 4.4. In particular, each interval in $\mathcal{P}_n$ is a *floor* in one of the two *buildings*.

In case when $R^n f \in \mathcal{D}^k_-$, renormalization cuts off a piece of the right tower and stacks it on top of the left tower to obtain the tower representation of $\mathcal{P}_{n+1}$, see Fig. 4.3.

In case when $R^n f \in \mathcal{D}^k_+$, renormalization cuts off a piece of the left tower and stacks it on top of the right tower to obtain the tower representation of $\mathcal{P}_{n+1}$, see Fig. 4.4.

Lemma 4.2.5 *Let $n \in \mathbb{N}$ and let $f \in \mathcal{D}^k$ n-renormalizable. If $R^n f$ is not renormalizable, i.e. $R^n f \in \mathcal{D}^k_0$, then all orbits of f converge to periodic orbits of the same period. In particular, if f is ∞-renormalizable then f is n-renormalizable for all $n \in \mathbb{N}$.*

Proof Let $g = R^n f \in \mathcal{D}^k_0$. We will first show that all orbits of g converge to a period-2 orbit.[1] By definition of the space $\mathcal{D}^k_0$ we have $c_g = v_g$. This implies

$$g : [0, c_g) \to [c_g, 1)$$

and

$$g : [c_g, 1) \to [0, c_g).$$

In particular, g has no fixed points. We denote $h = g^2$, then h is an increasing homeomorphism. Hence,

$$x_1 < x_2 \Rightarrow h(x_1) < h(x_2), \ \forall x_1, x_2 \in [0, 1).$$

For any $x \in [0, 1)$, if $h(x) > x$ then from the above observation we get

$$h^2(x) = h(h(x)) > h(x),$$

[1] One could prove this alternatively by first showing the rotation number of g is $1/2$ using the fact that 0 is a 2-periodic orbit, and use Exercise 2.5.9.

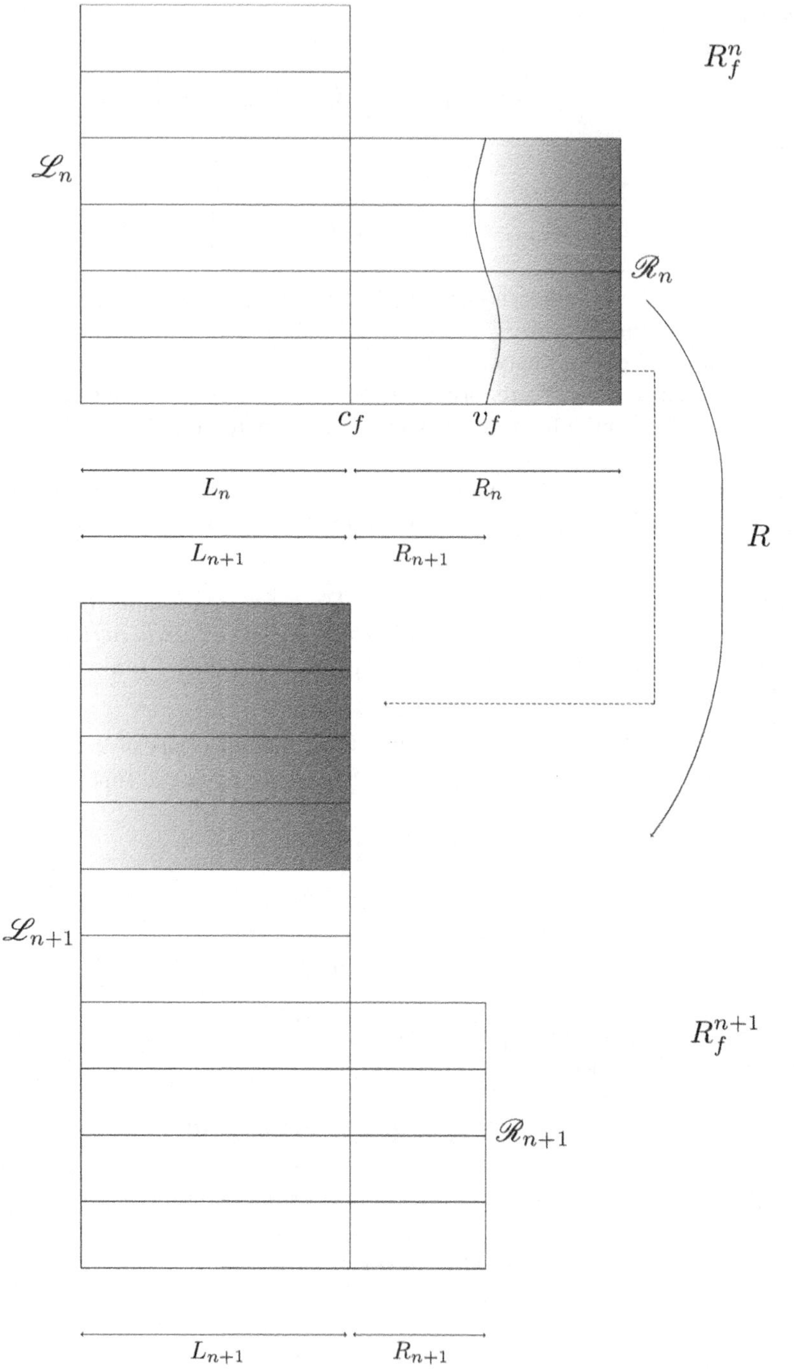

Fig. 4.3 The pre-renormalization of $f \in \mathcal{D}_-^k$

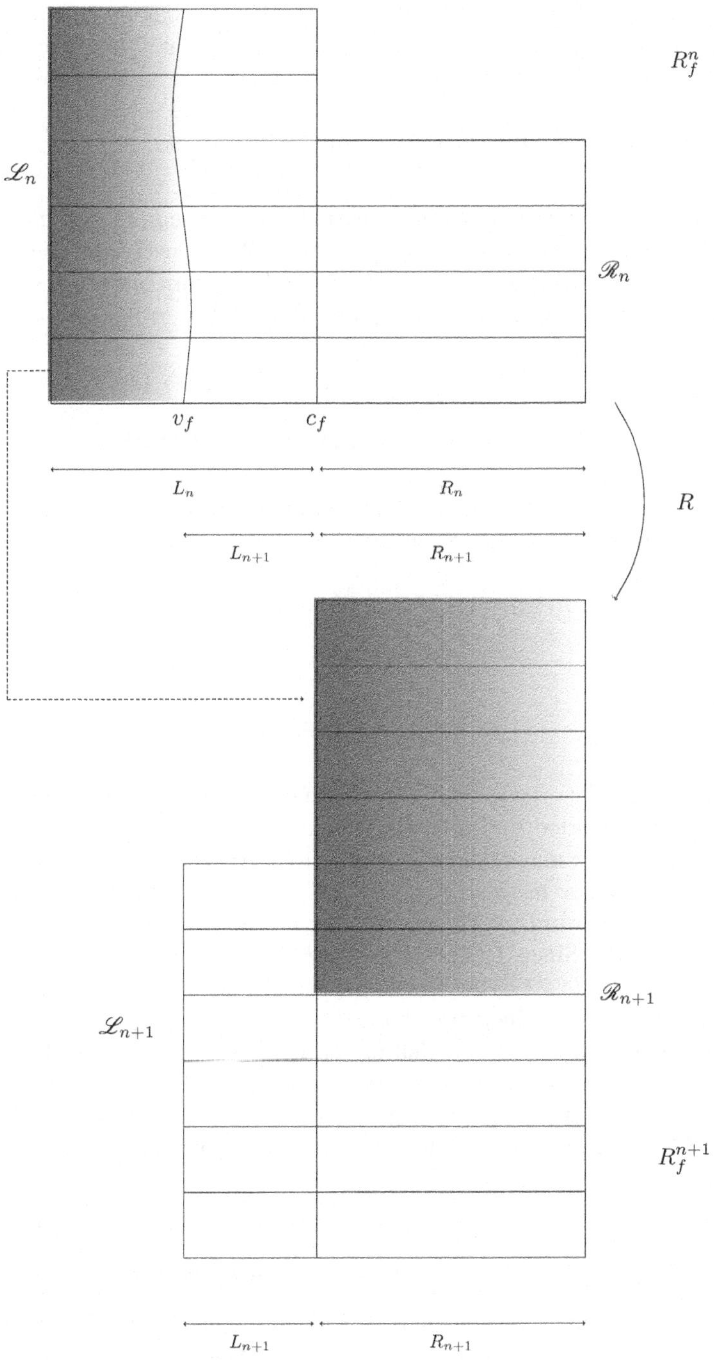

Fig. 4.4 The tower $\mathscr{P}_n$ when $R^n f \in \mathcal{D}_+^k$

similarly $h^3(x) > h^2(x)$, etc. Hence $\{h^m(x)\}_{m\in\mathbb{N}}$ is a monotone increasing sequence, it converges to a point $y \in [0,1]$. Notice that y is a fixed point of h since

$$h(y) = h(\lim_{m\to\infty} h^m(x)) = \lim_{m\to\infty} h^{m+1}(x) = y, \tag{4.2.1}$$

i.e. $g^2(y) = y$, y is a periodic point of period 2. The case $h(x) < x$ can be treated similarly. Hence we have proved that all orbits of $R^n f$ converge to a period-2 orbit.

By Lemmas 4.1.5 and 4.1.6, $g = R^n f$ is the first return map $F_n(x) : U_n \to U_n$ up to scaling, i.e. the two branches of g are (recall that $c_g = v_g$)

$$g_-(x) = (A_{U_n}^{-1} \circ f^{l_n} \circ A_{U_n})(x) : [0, c_g) \to [c_g, 1),$$

$$g_+(x) = (A_{U_n}^{-1} \circ f^{r_n} \circ A_{U_n})(x) : [c_g, 1) \to [0, c_g).$$

Hence $h(x) = (A_{U_n}^{-1} \circ f^{l_n+r_n} \circ A_{U_n})(x)$. For any point $x_0 \in [0,1)$, by the previous paragraph its orbit $\{h^m(x_0)\}_{m\in\mathbb{N}}$ converges to a fixed point $y_0 \in [0,1)$. That is,

$$h(y_0) = (A_{U_n}^{-1} \circ f^{l_n+r_n} \circ A_{U_n})(y_0) = y_0.$$

Hence

$$f^{l_n+r_n}(A_{U_n}(y_0)) = A_{U_n}(y_0),$$

i.e., every orbit $\{f^m(x)\}_{m\in\mathbb{N}}$ with $x \in U_n$ converges to the periodic orbit of $y' := A_{U_n}(y_0)$, and the period $k \le l_n + r_n$. We claim that $k = l_n + r_n$. In fact, without loss of generality we may assume $y' \in A_{U_n}([0, c_g))$. Then since $f^{l_n}(y')$ is the first return of y' to U_n, we have $y', f(y'), \ldots, f^{l_n-1}(y')$ are all distinct (otherwise the orbit returned earlier to U_n). The point $f^{l_n}(y') \in A_{U_n}([c_g, 1))$ is also distinct from these $l_n - 1$ points. Since $f^{r_n}(f^{l_n}(y'))$ is the first return of $f^{l_n}(y')$ to U_n, we have $f^{l_n+1}(y'), \ldots, f^{l_n+r_n-1}(y')$ are all distinct and distinct from $y', f(y'), \ldots, f^{l_n}(y')$. Hence the period of y' is indeed $l_n + r_n$.

From Lemma 4.2.3 we know that the orbit of every point in $[0,1)$ enters after some time in U_n. In particular, for every $x_0 \in [0,1)$, $f^k(x_0) \in U_n$ for some $k \in \mathbb{N}$. Thus the orbit $\{f^m(x_0)\}_{m\in\mathbb{N}}$ converges to the same periodic orbit as the orbit of $f^k(x_0)$. Hence we have proved that all orbits of f converge to periodic orbits of period $l_n + r_n$. □

Let $\mathscr{P}$ be a finite collection of intervals. We define $\text{mesh}(\mathscr{P}) = \max_{I\in\mathscr{P}} |I|$.

Lemma 4.2.6 *Let $f \in \mathcal{I}^k$, then mesh$(\mathscr{P}_n) \to 0$ when $n \to \infty$.*

Proof Fix $n \in \mathbb{N}$ and let $m_n = \max\{l_n, r_n\}$. By points 1. and 2. of Lemma 4.2.1 and Lemma 4.2.2 we know that m_n tends to infinity. Let us consider now the orbit of c_f under f until the time m_n, $O_m = \left\{c_f^i = f^i(c_f) | 0 \le i \le m_n\right\}$ and the partition generated by them $\mathscr{M}_n = \{M_n^i\}_{0 \le i \le m_n}$ where for all i, M_n^i is a connected component of $[0, 1] \setminus O_m$. Observe that if $I \in \mathscr{P}_n$, then $O_m \cap \mathring{I} = \emptyset$. From that,

$$\text{mesh}(\mathscr{P}_n) \le \text{mesh}(\mathscr{M}_n).$$

Since f does not have periodic orbits, by Proposition 2.2.6 we know that $\rho(f) \notin \mathbb{Q}$. By Theorem 3.3.1 and the fact that f is a piecewise C^k diffeomorphism with $k \ge 2$, f is conjugated to the rotation. Using Corollary 3.1.3 it follows that all orbits are dense, which implies that $\text{mesh}(\mathscr{M}_n) \to 0$. Hence

$$\text{mesh}(\mathscr{P}_n) \le \text{mesh}(\mathscr{M}_n) \to 0.$$

□

4.3 Renormalization Type

In this section we define a sequence of numbers which will characterize the topology of the circle map. The sequence counts the consecutive times that the renormalizations of the map are in $\mathcal{D}_+^k$ or in $\mathcal{D}_-^k$. The formal definition is as follows.

Let $f \in \mathcal{I}^k$. For all $n \in \mathbb{N}$ define $\sigma_n(f) \in \{+, -\}$ be such that

$$R^n f \in \mathcal{D}^k_{\sigma_n(f)}.$$

Let $\sigma(f) = \{\sigma_n(f)\}_{n \in \mathbb{N}} \in \{+, -\}^{\mathbb{N}}$. We define now a sequence of natural numbers $a(f) = \{a_n(f)\}_{n \ge 1}$ in the inductive way:

- $a_1(f) = \max\{i \ge 0 | \sigma_j(f) = +, \ \forall j \le i\}$,
- For $n \ge 2$,

$$a_n(f) = \max\{i \ge 0 | \sigma_{s_{n-1}+j}(f) = \sigma_{s_{n-1}+1}(f), \ \forall\, 0 < j \le i\},$$

where $s_{n-1} = \sum_{k=1}^{n-1} a_k(f)$.

The sequence $a(f)$ is called the renormalization type of f. The proof of the following remark is left as an exercise, see Exercise 4.7.8.

Remark 4.3.1 The definition of ∞-renormaliazability says that f does not have periodic orbits. This implies that $a_n(f)$ is a finite number for all $n \ge 1$.

The next proposition is illustrated in (Fig. 4.5).

Proposition 4.3.2 *Let $f, \tilde{f} \in \mathcal{I}^k$ with $a(f) = a(\tilde{f})$. Then for every $n \ge 0$ there exists a homeomorphism $h_n : [0, 1] \to [0, 1]$ such that*

1. *$h_n(\mathscr{P}_n) = \tilde{\mathscr{P}}_n$,*
2. *$h_n : f^i(L_n) \to \tilde{f}^i(\tilde{L}_n)$ is affine and onto for $0 \le i < l_n$,*

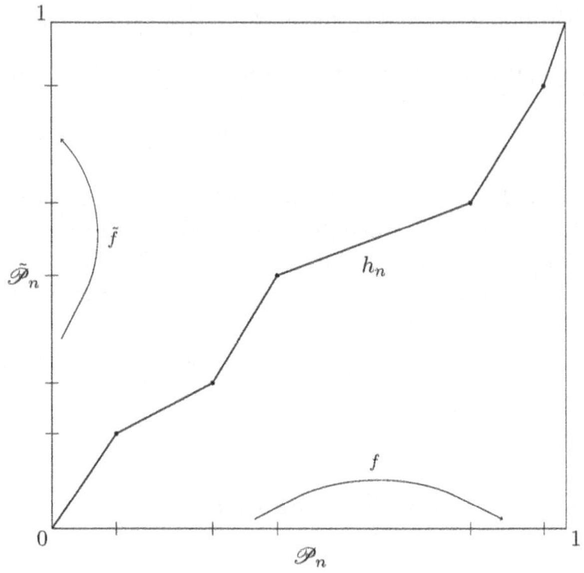

Fig. 4.5 The homeomorphism of Proposition 4.3.2

3. *$h_n : f^i(R_n) \to \tilde{f}^i(\tilde{R}_n)$ is affine and onto for $0 \le i < r_n$,*
4. *if $I \in \mathscr{P}_n$, then $h_{n+1}(I) = h_n(I)$.*

Proof Observe that the hypothesis $a(f) = a(\tilde{f})$, implies that $\sigma(f) = \sigma(\tilde{f})$. Lemmas 4.2.1 and 4.2.2 imply that for all $n \in \mathbb{N}$, $l_n = \tilde{l}_n$ and $r_n = \tilde{r}_n$.

The proof of the proposition is by induction. For $n = 0$, there is a unique orientation preserving piecewise affine homeomorphism $h_0 : [0, 1] \to [0, 1]$, with the properties that $h_0 : [0, c_f] \to [0, c_{\tilde{f}}]$ and $h_0 : [c_f, 1] \to [c_{\tilde{f}}, 1]$ are affine. Let us now proceed with the induction step. For all $n \in \mathbb{N}$ let $w_n \in [0, 1)$ a boundary point of R_n, namely $R_n = [c_f, w_n)$. Similarly define $\tilde{w}_n \in [0, 1)$ such that $\tilde{R}_n = [c_{\tilde{f}}, \tilde{w}_n)$. Suppose that $R^n f \in \mathcal{D}^k_-$. By Lemma 4.2.1 we know that $R_{n+1} = R_n \setminus f^{l_n}(L_n) = [c_f, w_{n+1})$. Similarly $\tilde{R}_{n+1} = \tilde{R}_n \setminus \tilde{f}^{l_n}(\tilde{L}_n) = [c_{\tilde{f}}, \tilde{w}_{n+1})$. We define now h_{n+1}. By Remark 4.2.4, for $0 \le i < l_n$, $f^i(L_n) \in \mathscr{P}_{n+1}$. As consequence we define

$$h_{n+1}|_{f^i(L_n)} = h_n.$$

For $0 \le i < r_n$, consider the point $w^i_{n+1} \in f^i(R_n) \in \mathscr{P}_n$ where $w^i_{n+1} = f^i(w_{n+1})$. Similarly define $\tilde{w}^i_{n+1} \in \tilde{f}^i(\tilde{R}_n) \in \tilde{\mathscr{P}}_n$. According to Lemma 4.2.1, $f^i(R_n) = f^i([c_f, w_{n+1})) \cup f^i([w_{n+1}, w_n))$ and similarly $\tilde{f}^i(\tilde{R}_n) = \tilde{f}^i([c_{\tilde{f}}, \tilde{w}_{n+1})) \cup \tilde{f}^i([\tilde{w}_{n+1}, \tilde{w}_n))$. Define a piecewise affine map $h_{n+1}|_{f^i(R_n)}$ with the properties that

$$h_{n+1}|_{f^i(R_n)} : f^i([c_f, w_{n+1})) \to \tilde{f}^i([c_{\tilde{f}}, \tilde{w}_{n+1}))$$

and

$$h_{n+1}|_{f^i(R_n)} : f^i([w_{n+1}, w_n)) \to \tilde{f}^i([\tilde{w}_{n+1}, \tilde{w}_n))$$

are orientation preserving affine homeomorphisms. Observe that $h_{n+1}|_{\partial f^i(R_n)} = h_n$. Consequently h_{n+1} is an homeomorphism and it satisfies all the properties of the lemma. The case $R^n f \in \mathcal{D}_+^k$ is a repetition of the previous arguments. □

Proposition 4.3.3 *Let $f, \tilde{f} \in \mathcal{I}^k$ with $a(f) = a(\tilde{f})$. Then $h_n : [0, 1] \to [0, 1]$ converges to an homeomorphism h which is a conjugacy, $h \circ f = \tilde{f} \circ h$.*

Proof Let $I \in \mathscr{P}_n$, then from point 4. of Proposition 4.3.2 we have that for all $m > 0$

$$h_{n+m}(I) \subset h_n(I).$$

Hence that for every $x \in I$,

$$|h_{n+m}(x) - h_n(x)| \leq |h_n(I)| \leq mesh(\tilde{\mathscr{P}}_n).$$

By Lemma 4.2.6 it follows that h_n is a Cauchy sequence in the uniform topology. Hence h_n converges to a continuous function h. For proving that h is an homeomorphism, apply the same argument to h_n^{-1}.

It remains to prove that, given $x \in [0, 1]$, $h(f(x)) = \tilde{f}(h(x))$. For all $n \in N$, let I_n denote the unique interval in $\mathscr{P}_n$ containing x. The existence of such I_n for each $n \in \mathbb{N}$ is ensured by Lemma 4.2.3. By Lemma 4.2.6, $I_n \to x$ and $h(I_n) \to h(x)$, hence it suffices to show that $h(f(I_n)) = \tilde{f}(h(I_n))$ for all n. Given any interval $I \in \mathscr{P}_n$, by property 4 of Proposition 4.3.2, $h_m(I) = h_n(I)$ for all $m \geq n$, hence we have

$$h(I) = h_n(I), \quad \forall I \in \mathscr{P}_n, n \in \mathbb{N}. \tag{4.3.1}$$

For any fixed $n \in \mathbb{N}$, without loss of generality we may assume $I_n = f^i(R_n)$ for some $0 \leq i < r_n$, then $h_n(I_n) = \tilde{f}^i(\tilde{R}_n)$ by property 3 of Proposition 4.3.2. Using (4.3.1) we get

$$\tilde{f}(h(I_n)) = \tilde{f}(h_n(I_n)) = \tilde{f}(\tilde{f}^i(\tilde{R}_n)) = \tilde{f}^{i+1}(\tilde{R}_n). \tag{4.3.2}$$

For $h(f(I_n)) = h(f^{i+1}(R_n))$, if $i < r_n - 1$ then we have $f^{i+1}(R_n) \in \mathscr{P}_n$, so

$$h(f(I_n)) = h_n(f^{i+1}(R_n)) = \tilde{f}^{i+1}(\tilde{R}_n), \quad \forall\, 0 \leq i < r_n - 1,$$

where we again used (4.3.1) and Proposition 4.3.2. Now suppose that $i = r_n - 1$, namely $f(I_n) = f^{r_n}(R_n)$. If $R^n f \in \mathcal{D}_+^k$, by Lemma 4.2.2 we have $R_{n+1} = R_n$ and $r_{n+1} = r_n + l_n$. In particular $f(I_n) = f^{r_n}(R_{n+1})$ and $r_n \leq r_{n+1} - 1$, which implies that in this case $f(I_n) \in \mathscr{P}_{n+1}$. Since $a(f) = a(\tilde{f})$ we have that $R^n \tilde{f}$ is also in $\mathcal{D}^k_+$. Hence

$$h(f(I_n)) = h_{n+1}(f^{r_n}(R_{n+1})) = \tilde{f}^{r_n}(\tilde{R}_{n+1}) = \tilde{f}^{r_n}(\tilde{R}_n)$$
$$= \tilde{f}^{i+1}(\tilde{R}_n),\ i = r_n - 1 \text{ and } f \in \mathcal{D}^k_+.$$

If $R^n f \in \mathcal{D}^k_-$, note that $f^{r_n}(R_n) = U_{n+1} = L_{n+1} \sqcup R_{n+1}$. Similarly as in the case of $\mathcal{D}^k_+$ we have that

$$h(f(I_n)) = h(L_{n+1} \cup R_{n+1}) = h(L_{n+1}) \cup h(R_{n+1})$$
$$= \tilde{L}_{n+1} \cup \tilde{R}_{n+1} = \tilde{U}_{n+1} = \tilde{f}^{r_n}(\tilde{R}_n) = \tilde{f}^{i+1}(\tilde{R}_n), \quad i = r_n - 1 \text{ and } f \in \mathcal{D}^k_-.$$

We have proved that $h(f(I_n) = \tilde{f}^{i+1}(\tilde{R}_n)$. Together with (4.3.2) we conclude that $\tilde{f}(h(I_n)) = h(f(I_n))$ for all $n \in \mathbb{N}$. □

The proof of the following theorem is left as an exercise, see Exercise 4.7.13.

Theorem 4.3.4 *Let $f, \tilde{f} \in \mathcal{I}^k$. The maps f and $\tilde{f}$ are topologically conjugated if and only if $a(f) = a(\tilde{f})$.*

4.4 Two Topological Invariants

In the previous sections we proved that to any ∞-renormalizable (i.e. maps without periodic orbits) generalized circle diffeomorphism f, it is possible to associate an irrational rotation number $\rho(f)$ and a renormalization type $a(f)$. We proved that both $\rho(f)$ and $a(f)$ are topological invariant. The aim of this section is to explain the relation between these two invariants.

4.4.1 Continued Fractions

Given a sequence of integers $a_n \geq 1$, $n \in \mathbb{N}$, define the corresponding fractions

$$\cfrac{1}{a_1 + \cfrac{1}{a_2 + \cfrac{1}{\ddots + \cfrac{1}{a_n}}}} = [a_1, a_2, \ldots, a_n].$$

Lemma 4.4.1 *Given a sequence of integers $a_1, a_2, \ldots$, if the sequences $\{p_n\}_{n\in\mathbb{N}}$ and $\{q_n\}_{n\in\mathbb{N}}$ satisfy the following recursion formulas:*

$$\begin{cases} p_0 = 0, q_0 = 1, p_1 = 1, q_1 = a_1, & \\ p_{n+1} = a_{n+1}p_n + p_{n-1} & n \geq 1 \\ q_{n+1} = a_{n+1}q_n + q_{n-1} & n \geq 1 \end{cases} \tag{4.4.1}$$

then

$$[a_1, a_2, \ldots, a_n] = \frac{p_n}{q_n}.$$

Proof Suppose sequences of numbers $\{p_n\}$, $\{q_n\}$ are given by (4.4.1). By induction we prove that for all $n \geq 1$ and $t \geq 0$,

$$[a_1, \ldots, a_n + t] = \frac{p_n + tp_{n-1}}{q_n + tq_{n-1}}.$$

For $n = 1$

$$[a_1 + t] = \frac{1}{a_1 + t} = \frac{p_1 + tp_0}{q_1 + tq_0}.$$

Supposing that this is true for n we prove it for $n + 1$.

$$\begin{aligned} [a_1, \ldots, a_n, a_{n+1} + t] &= \left[a_1, \ldots, a_n + \frac{1}{a_{n+1} + t}\right] \\ &= \frac{p_n + \left(\frac{1}{a_{n+1}+t}\right) p_{n-1}}{q_n + \left(\frac{1}{a_{n+1}+t}\right) q_{n-1}} \\ &= \frac{a_{n+1}p_n + tp_n + p_{n-1}}{a_{n+1}q_n + tq_n + q_{n-1}} \\ &= \frac{p_{n+1} + tp_n}{q_{n+1} + tq_n} \end{aligned}$$

where $p_{n+1} = a_{n+1}p_n + p_{n-1}$ and $q_{n+1} = a_{n+1}q_n + q_{n-1}$ by (4.4.1). This finishes the induction step. Now the lemma follows by taking $t = 0$. □

The two sequences of numbers $\{p_n\}_{n\in\mathbb{N}}$ and $\{q_n\}_{n\in\mathbb{N}}$ are called the convergents of the continued fraction. It turns out that $\lim_{n\to\infty} \frac{p_n}{q_n}$ always exists, this will be a consequence of the next theorem, Theorem 4.4.2. For a further discussion on the convergence of continued fractions see Sect. 4.4.3. If

$$\lim_{n\to\infty} \frac{p_n}{q_n} = \alpha$$

then we write

$$\alpha = \cfrac{1}{a_1 + \cfrac{1}{a_2 + \cfrac{1}{\ddots}}} = [a_1, a_2, \dots].$$

The continued fraction expansion of α is $[a_1, a_2, \dots]$.

4.4.2 First Return Times

In this subsection we prove that the renormalization type of a map determines the rotation number as a continued fraction. The two topological invariants, $a(f)$ and $\rho(f)$ are equivalent. The first return times of $R^n f$, see Lemma 4.1.6, determine the connection.

Theorem 4.4.2 *Let $f \in \mathcal{I}^k$ with rotation number $\rho(f)$ and renormalization type $a(f)$, then*

$$\rho(f) = \cfrac{1}{1 + \cfrac{1}{[a(f)]}} = [\tilde{a}_1, \tilde{a}_2, \tilde{a}_3, \dots],$$

where $\tilde{a}_1 = a_1(f) + 1$, and $\tilde{a}_n = a_n(f)$ for all $n \geq 2$.

Proof Let l_n and r_n be the first return times of the nth-renormalization. Let $s_0 = 0$ and for $n \geq 1$ let $s_n = \sum_{k=1}^{n} a_k(f)$. We define

$$p_n^- = \#\left\{0 \leq i < l_n | f^i(L_n) \subset [c_f, 1]\right\}$$

and

$$p_n^+ = \#\left\{0 \leq i < r_n | f^i(R_n) \subset [c_f, 1]\right\}.$$

By Lemma 2.1.6 and the definition of rotation number (namely Proposition 2.2.1) we have

$$\rho(f) = \lim_{n\to\infty} \frac{1}{l_n}(F^{l_n}(x) - x) = \lim_{n\to\infty} \frac{1}{l_n}(p_n^- + \tilde{f}^{l_n}(x) - x) = \lim_{n\to\infty} \frac{p_n^-}{l_n},$$

where $x \in L_n$, F and $\tilde{f}$ are the corresponding lift and interval map of f as in Lemma 2.1.6. Similarly, choosing $x \in R_n$ we have

$$\rho(f) = \lim_{n\to\infty} \frac{1}{r_n}(F^{r_n}(x) - x) = \lim_{n\to\infty} \frac{1}{r_n}(p_n^+ + \tilde{f}^{r_n}(x) - x) = \lim_{n\to\infty} \frac{p_n^+}{r_n}.$$

Hence we have proved

$$\frac{p_n^-}{l_n}, \frac{p_n^+}{r_n} \to \rho(f). \tag{4.4.2}$$

Let for $n \in \mathbb{N}$

$$q_{2n} = l_{s_{2n}}, \; p_{2n} = p_{s_{2n}}^-$$

and

$$q_{2n+1} = r_{s_{2n+1}}, \; p_{2n+1} = p_{s_{2n+1}}^+.$$

From the definition we have $q_0 = 1$, $p_0 = 0$, $p_1 = 1$ and from Lemma 4.2.2 we know that $q_1 = a_1(f) + 1 = \tilde{a}_1$. In particular the sequences $\{p_n\}_{n\in\mathbb{N}}$ and $\{q_n\}_{n\in\mathbb{N}}$ satisfy the initial conditions of (4.4.1). We prove now that they actually satisfy the complete recursion formulas.

Observe that for $n \geq 0$ from $R^{s_{2n+1}} f$ to $R^{s_{2n+2}} f$ we have to do $a_{2n+2}(f) = \tilde{a}_{2n+2}$ renormalization steps in $\mathcal{D}_-^k$. From Lemma 4.2.1 we get:

1^-: $q_{2n+2} = r_{s_{2n+1}} + \tilde{a}_{2n+2} q_{2n+1}$,
2^-: $p_{2n+2} = p_{s_{2n+1}}^- + \tilde{a}_{2n+2} p_{2n+1}$,
3^-: $r_{s_{2n+2}} = r_{s_{2n+1}} = q_{2n+1}$,
4^-: $p_{s_{2n+2}}^+ = p_{s_{2n+1}}^+ = p_{2n+1}$.

Observe that for $n \geq 1$ from $R^{s_{2n}} f$ to $R^{s_{2n+1}} f$ we have to do $a_{2n+1}(f) = \tilde{a}_{2n+1}$ renormalization steps in $\mathcal{D}_+^k$. From Lemma 4.2.2 we get:

1^+: $q_{2n+1} = r_{s_{2n}} + \tilde{a}_{2n+1} q_{2n}$,
2^+: $p_{2n+1} = p_{s_{2n}}^+ + \tilde{a}_{2n+1} p_{2n}$,
3^+: $l_{s_{2n+1}} = l_{s_{2n}} = q_{2n}$,
4^+: $p_{s_{2n+1}}^- = p_{s_{2n}}^- = p_{2n}$.

Notice that:

$$1^+ \text{ and } 3^- \implies q_{2n+1} = \tilde{a}_{2n+1} q_{2n} + q_{2n-1},$$

$$2^+ \text{ and } 4^- \implies p_{2n+1} = \tilde{a}_{2n+1} p_{2n} + p_{2n-1},$$

$$1^- \text{ and } 3^+ \implies q_{2n+2} = \tilde{a}_{2n+2} q_{2n+1} + q_{2n},$$

$$2^- \text{ and } 4^+ \implies p_{2n+2} = \tilde{a}_{2n+2} p_{2n+1} + p_{2n}.$$

Hence the sequences $\{p_n\}_{n\in\mathbb{N}}$ and $\{q_n\}_{n\in\mathbb{N}}$ satisfy the recursion formulas in (4.4.1) with $a_n = \tilde{a}_n$ for all n. Consequently, by Lemma 4.4.1 and (4.4.2) we get

$$\rho(f) = \lim_{n\to\infty} \frac{p_n}{q_n} = [\tilde{a}_1, \tilde{a}_2, \dots].$$

This concludes the proof. □

4.4.3 Continued Fractions and Closest Approaches

For any $\rho \in [0, 1)\backslash\mathbb{Q}$, let r_ρ the rotation of angle ρ, then r_ρ has rotation number ρ and it is ∞-renormalizable. Hence $\rho = [a_1(r_\rho) + 1, a_2(r_\rho), a_3(r_\rho), \dots]$. In particular, Theorems 4.3.4 and 4.4.2 imply that every irrational number can be represented in a unique way by a continued fraction. On the other hand, Exercise 4.7.10 and Theorem 4.4.2 imply that for every given sequence of integers $a_1, a_2, \dots,$ the continued fraction $[a_1, a_2, \dots]$ converges. We proved this classical theorem by identifying the times of closest approach, l_n, and r_n from Lemma 4.1.6, with the convergents q_n of the continued fraction corresponding with the renormalization type. This is the essence of the proof of Theorem 4.4.2.

Of course, this theorem in number theory on continued fraction is normally obtained without using dynamics. It uses the Gauß-map $G : (0, 1) \longrightarrow [0, 1)$ which is defined as

$$G(x) = \frac{1}{x} - \left\lfloor \frac{1}{x} \right\rfloor, \forall x \in (0, 1),$$

where $\lfloor z \rfloor$ denotes the integer part of z. Given an irrational $\alpha \in (0, 1)$. The iterates $G(\alpha), \dots, G^{n-1}(\alpha), \dots$ are different from zero and one considers the natural numbers

$$a_1 = \left\lfloor \frac{1}{\alpha} \right\rfloor, a_2 = \left\lfloor \frac{1}{G(\alpha)} \right\rfloor, \dots, a_n = \left\lfloor \frac{1}{G^{n-1}(\alpha)} \right\rfloor, \cdots$$

Each irrational number $\alpha \in (0, 1)$ can be represented uniquely as continued fraction. Namely,

$$\alpha = [a_1, a_2, \dots],$$

A traditional discussion of continued fractions can be found in [HW08].

The rational numbers

$$\frac{p_n}{q_n} = [a_1, a_2, \cdots, a_n]$$

are characterized by the fact that they are the best approximations of ρ. Namely, for all $p, q \leq q_n$ we have

$$|p - q\alpha| \geq |p_n - q_n\alpha|.$$

Observe that this inequality corresponds exactly to the closest approaches as discussed in Lemmas 3.3.2 and 4.1.6.

The approximations of irrational numbers by rational numbers is very well understood. The Gauß-map gives an algorithm to find the best approximations. In higher dimensions the approximation of vectors by rational vectors is not as well understood.

4.5 Renormalization of Rotations

The set of rotations

$$\mathcal{W} = \{r_\rho | \rho \in [0, 1)\} \subset \mathcal{D}^k$$

plays a special role. It will be shown in the next chapter that $\mathcal{W}$ is the attractor of renormalization, namely for all $f \in \mathcal{I}^k$, $R^n f \to \mathcal{W}$. In this section we describe renormalization of rotations. We identify the rotations with their rotation number. In particular, $\mathcal{W} \subset \mathcal{D}^k$ is a curve.

Lemma 4.5.1 *The set of rotations $\mathcal{W}$ is invariant under renormalization. In particular $R : \mathcal{W} \backslash \{r_{\frac{1}{2}}\} \to \mathcal{W}$ and*

$$R : \rho \mapsto \frac{\rho}{1 - \rho} \text{ when } \rho < \frac{1}{2},$$

and

$$R : \rho \mapsto \frac{2\rho - 1}{\rho} \text{ when } \rho > \frac{1}{2}.$$

Moreover, for every $\rho \in (0, \frac{1}{2}) \cup (\frac{1}{2}, 1)$

$$\frac{\partial R}{\partial \rho}(\rho) > 1.$$

4.6 Distortion Bounds

In the next chapter we will show the convergence of renormalization. The first step towards this convergence is the following lemma, which states that the renormalizations stay in a bounded set. This is know as the *a priori bounds*, they play a central role in renormalization. Precisely, we define the distortion for a generalized circle diffeomorphism $f \in \mathcal{D}^k$ as

$$\mathrm{Dist}(f) = \mathrm{Dist}(f_-) + \mathrm{Dist}(f_+),$$

then the renormalizations of ∞-renormalizable maps are bounded s stated in the following lemma.

Lemma 4.6.1 *For all $f \in \mathcal{I}^k$ and for all $n \in \mathbb{N}$,*

$$\mathrm{Dist}(R^n f) \leq \int |\eta_f| < \infty.$$

In particular,

- $\mathrm{Dist}(f^i|_{L_n}) \leq \int |\eta_f| < \infty$ *for* $0 \leq i \leq l_n$,
- $\mathrm{Dist}(f^i|_{R_n}) \leq \int |\eta_f| < \infty$ *for* $0 \leq i \leq r_n$.

The proof of the lemma is analogous to the proof of Proposition 3.2.4. We adapt the argument to the setting of renormalization and present the proof here.

Proof From Lemma 4.1.6 we know that the nth pre-renormalization of f has the form:

$$pR^n f(x) = \begin{cases} f^{l_n}(x) & x \in L_n \\ f^{r_n}(x) & x \in R_n \end{cases}$$

We estimate $\mathrm{Dist}(pR^n f|_{L_n})$. Let $x, y \in L_n$, then by Proposition 3.2.4,

$$\log \frac{DpR^n f(x)}{DpR^n f(y)} = \log \frac{Df^{l_n}(x)}{Df^{l_n}(y)} = \sum_{i=0}^{l_n-1} \log \frac{Df(f^i(x))}{Df(f^i(y))}$$

$$\leq \sum_{i=0}^{l_n-1} \int_{f^i(L_n)} |\eta_f| = \sum_{I \in \mathcal{L}_n} \int_I |\eta_f|.$$

On R_n, let $x, y \in R_n$. Similarly as before we get:

$$\log \frac{DpR^n f(x)}{DpR^n f(y)} \leq \sum_{I \in \mathcal{R}_n} \int_I |\eta_f|.$$

Recall that $R^n f$ differs from $pR^n f$ by an affine rescaling which has zero distortion, then Lemma 3.2.3 implies $\mathrm{Dist}(R^n f) = \mathrm{Dist}(pR^n f)$. Hence

$$\mathrm{Dist}(R^n f) \leq \sum_{I \in \mathscr{P}_n} \int_I |\eta_f| = \int |\eta_f|.$$

□

4.7 Exercises

Exercise 4.7.1 Prove that for each $k \geq 2$, $\mathcal{D}^k$ is path connected.

Exercise 4.7.2 Prove Remark 4.1.1.

Exercise 4.7.3 Prove Lemma 4.1.2.

Exercise 4.7.4 Show that the renormalization operator R defined in Definition 4.1.3 is well-defined, i.e., the pre-renormalization pRf is surjective, and Rf is indeed contained in the space $\mathcal{D}^k$.

Exercise 4.7.5 Prove that the renormalization operator R is a continuous map to $\mathcal{D}^k$.

Exercise 4.7.6 Prove Lemma 4.1.5.

Exercise 4.7.7 Prove Lemma 4.1.6.

Exercise 4.7.8 Prove Remark 4.3.1.

Exercise 4.7.9 Consider a map which has renormalization type $(+, +, -, +)$. Sketch the dynamical partions $\mathscr{P}_1$, $\mathscr{P}_2$, $\mathscr{P}_3$, and $\mathscr{P}_4$.

Exercise 4.7.10 Given a sequence $\{\sigma_n\}_{n \in \mathbb{N}} \in \{+, -\}^{\mathbb{N}}$, construct a rotation $f \in \mathcal{W}$ such that $\sigma(f) = \{\sigma_n\}_{n \in \mathbb{N}}$. Hint: Use an inductive argument and Lemma 4.5.1.

Exercise 4.7.11 Consider a map which has renormalization type

$$\sigma = (+, +, -, +, +, -, +, +, -, \cdots).$$

Determine its rotation number. Hint: Show it is an algebraic number.

Exercise 4.7.12 Consider a map which has renormalization type

$$\sigma = (+, +, -, -, +, +, -, -, +, +, -, - \cdots).$$

Determine its rotation number.

Exercise 4.7.13 Prove Theorem 4.3.4.

Exercise 4.7.14 Prove Lemma 4.5.1.

Chapter 5
The Three Ingredients for Convergence of Renormalization

The renormalization operator was introduced to study the small scale properties of generalized circle diffeomorphisms. It turns out that on small scale, these maps behave like rotations. In the sense that consecutive renormalizations converge to rotations. This central and impressive aspect of renormalization theory implies that the geometrical aspects of orbits on small scales are determined by the rotations. The following three chapters are dedicated to explaining this convergence phenomenon. First, in this chapter, we discuss the three ingredients which are responsible for the convergence of renormalization to the rotations. There is a specific mechanism which is responsible for each of the following aspects of convergence.

1. Renormalizations of $\mathcal{C}^3$ generalized circle diffeomorphisms converge to Möbius circle diffeomorphisms.
2. Renormalizations of Möbius circle diffeomorphisms converge to affine circle diffeomorphisms.
3. Renormalizations of affine circle diffeomorphisms converge to rotations.

Some of these mechanisms might be used in other context.

5.1 From Diffeomorphisms to Möbius Maps

From this moment on we are going to work with functions in $\mathcal{D}^2$. Only in Lemma 5.1.10 the estimates need a $\mathcal{C}^3$ bound and this is the reason why the convergence theorem of the renormalizations require that the function is in $\mathcal{D}^3$.

We introduce now a more convenient distance in $\mathcal{D}^2$. First we denote by $\text{Diff}^2_+([0,1])$ the set of $\mathcal{C}^2$ diffeomorphisms of the unit interval and we define a distance on it as

$$dist_\eta(\varphi, \psi) = \int |\eta_\varphi - \eta_\psi|,$$

Y. Dong et al., *One-Dimensional Dynamics*, Lecture Notes of the Unione Matematica Italiana 30, https://doi.org/10.1007/978-3-031-94882-4_5

with $\varphi, \psi \in \text{Diff}^2_+([0, 1])$. Let $\phi : I \to J$ be a diffeomorphism between two intervals. Recall that the normalized version $N\phi : [0, 1] \to [0, 1]$ of ϕ is obtained by rescaling domain and range of ϕ,

$$N\phi = A_J^{-1} \circ \phi \circ A_I,$$

where $A_I : [0, 1] \to I$ and $A_J : [0, 1] \to J$ are the unique orientation preserving affine diffeomorphisms. Consequently we define the non-linearity distance for $f, g \in \mathcal{D}^2$ as,

$$d_\eta(f, g) = \max\left\{|c_f - c_g|, |v_f - v_g|, dist_\eta(Nf_-, Ng_-), dist_\eta(Nf_+, Ng_+)\right\}.$$

Lemma 5.1.1 *Let $D \subset \text{Diff}^2_+([0, 1])$ be a set for which there exists $L > 0$, such that $\sup_x |\eta_\varphi(x)| \leq L$ for every $\varphi \in D$. Then $dist_{C^1}$ is bounded by $dist_\eta$ on D, i.e. there exists constant $C > 0$ such that*

$$dist_{C^1}(\varphi, \psi) \leq C dist_\eta(\varphi, \psi), \quad \forall \varphi, \psi \in D.$$

In particular, let $B \subset \mathcal{D}^2$ be a set for which there exists $b > 0$ such that for every $f \in B$, $\sup_x |\eta_f(x)| \leq b$, then the distance d_1 is bounded by d_η on B.

Proof First notice that any $\varphi \in \text{Diff}^2_+([0, 1])$ and its first derivative $D\varphi$ are determined by its non-linearity η_φ. In fact, by definition we have $\eta_\varphi(x) = D \log D\varphi(x)$, hence

$$\log D\varphi(x) = \log D\varphi(0) + \int_0^x \eta_\varphi(t)dt, \quad \forall x \in [0, 1],$$

thus for all $x \in [0, 1]$,

$$D\varphi(x) = D\varphi(0)e^{\int_0^x \eta_\varphi(t)dt}.$$

Moreover, since $\varphi(0) = 0, \varphi(1) = 1$, we have $\int_0^1 D\varphi(x)dx = \varphi(1) - \varphi(0) = 1$, hence

$$\int_0^1 e^{\int_0^s \eta_\varphi(t)dt} ds = \frac{1}{D\varphi(0)}.$$

Using the previous two equalities we get

$$\varphi(x) = \int_0^x D\varphi(s)ds = D\varphi(0)\int_0^x e^{\int_0^s \eta_\varphi(t)dt} ds = \frac{\int_0^x e^{\int_0^s \eta_\varphi(t)dt} ds}{\int_0^1 e^{\int_0^s \eta_\varphi(t)dt} ds}, \tag{5.1.1}$$

$$D\varphi(x) = \frac{e^{\int_0^x \eta_\varphi(t)dt}}{\int_0^1 e^{\int_0^s \eta_\varphi(t)dt} ds}. \tag{5.1.2}$$

Suppose $\phi, \psi \in D$ are two diffeomorphisms. We need to estimate $|\phi - \psi|$ and $|D\phi - D\psi|$ in terms of $dist_\eta(\phi, \psi) = \int |\eta_\phi - \eta_\psi|$. For simplifying the notation let us denote

$$\int_0^s \eta_\phi(t)dt = \Phi(s), \int_0^s \eta_\psi(t)dt = \Psi(s).$$

Observe that

$$\left|\int_0^1 e^{\Phi(s)}ds\right| \geq \left|\int_0^1 e^{-Ls}ds\right| \geq \left|\int_0^1 e^{-L}ds\right| \geq e^{-L}.$$

Using this and (5.1.1) we have

$$\begin{aligned}
&|\phi(x) - \psi(x)| \\
&= \frac{1}{\left|\int_0^1 e^{\Phi(s)}ds \int_0^1 e^{\Psi(s)}ds\right|} \cdot \left| \int_0^x e^{\Phi(s)}ds\left(\int_0^1 e^{\Psi(s)}ds - \int_0^1 e^{\Phi(s)}ds\right)\right. \\
&\quad \left. + \int_0^1 e^{\Phi(s)}ds \left(\int_0^x e^{\Phi(s)}ds - \int_0^x e^{\Psi(s)}ds\right)\right| \\
&\leq e^{2L}\left|\int_0^x e^{\Phi(s)}ds \int_0^1 \left(e^{\Psi(s)} - e^{\Phi(s)}\right)ds + \int_0^1 e^{\Phi(s)}ds \int_0^x \left(e^{\Phi(s)} - e^{\Psi(s)}\right)ds\right|.
\end{aligned}$$

Note that since $|\Phi(s)| \leq \sup_x |\eta_\phi(x)| \leq L$, $|\Psi(s)| \leq \sup_x |\eta_\psi(x)| \leq L$, by the mean value theorem we have

$$\begin{aligned}
\left|e^{\Phi(s)} - e^{\Psi(s)}\right| &\leq e^L |\Phi(s) - \Psi(s)| \leq e^L \int_0^s |\eta_\phi(t) - \eta_\psi(t)|dt \\
&\leq e^L dist_\eta(\phi, \psi).
\end{aligned} \tag{5.1.3}$$

Hence

$$|\phi(x) - \psi(x)| \leq e^{2L} \cdot \left(e^L \cdot e^L dist_\eta(\phi, \psi) + e^L \cdot e^L dist_\eta(\phi, \psi)\right) = 2e^{4L} dist_\eta(\phi, \psi).$$

Applying similar arguments to $|D\phi - D\psi|$ we get

$$dist_{C^1}(\phi, \psi) = \max\{\sup|\phi - \psi|, \sup|D\phi - D\psi|\} \leq C dist_\eta(\phi, \psi)$$

for some constant C, and the proof is complete. □

Remark 5.1.2 As we computed in (5.1.3) in the above proof, the mean value theorem implies that the exponential function is Lipschitz on bounded intervals, namely for all $x, y \in [-K, K]$, $K > 0$, there exist two constants $C_1, C_2 > 0$, depending on K, such that

$$C_1|x-y| \leq |e^x - e^y| \leq C_2|x-y|.$$

Moreover, the same argument yields that the logarithm function is Lipschitz on $[\sigma, \infty)$ for any $\sigma > 0$. These facts will be frequently used in future discussions.

We define now a Möbius map.

Definition 5.1.3 Let $w \in \mathbb{R}$. The Möbius map of weight w, $m_w : [0,1] \to [0,1]$ is defines as

$$m_w(x) = \frac{x}{(1-e^{\frac{w}{2}})x + e^{\frac{w}{2}}}.$$

The following lemma states a collection of basic properties of Möbius map. The proof is done by direct computation and is left to the reader (see Exercise 5.4.1).

Lemma 5.1.4 *For all $w \in \mathbb{R}$, the Möbius map m_w satisfies:*

$$\int \eta_{m_w} = \log \frac{Dm_w(1)}{Dm_w(0)} = w, \tag{5.1.4}$$

$$Dm_w(x) = \frac{e^{\frac{w}{2}}}{[(1-e^{\frac{w}{2}})x + e^{\frac{w}{2}}]^2} \quad \forall x \in [0,1], \tag{5.1.5}$$

$$D^2 m_w(x) = \frac{2e^{\frac{w}{2}}(e^{\frac{w}{2}}-1)}{[e^{\frac{w}{2}}(1-x)+x]^3}, \tag{5.1.6}$$

and

$$\eta_{m_w}(x) = \frac{2(e^{\frac{w}{2}}-1)}{e^{\frac{w}{2}}(1-x)+x}. \tag{5.1.7}$$

Moreover, m_w is an analytic strictly increasing diffeomorphism.

Lemma 5.1.5 *Let $0 \leq a < b \leq 1$, and m be the Möbius map of weight w. Then $Nm|_{[a,b]}$ is again a Möbius map of weight $w' = \log(Dm(b)/Dm(a))$. In particular, if $a = 0$ and $b = \sigma \in [0,1]$, then*

$$(e^{-\frac{w'}{2}} - 1) = \sigma(e^{-\frac{w}{2}} - 1).$$

Proof Since by Lemma 5.1.4 the Möbius map is strictly increasing, by direct computation we have

$$A_{m[a,b]}^{-1}(x) = \frac{[(1-e^{\frac{w}{2}})a + e^{\frac{w}{2}}][(1-e^{\frac{w}{2}})b + e^{\frac{w}{2}}]}{e^{\frac{w}{2}}(b-a)} \left(x - \frac{a}{(1-e^{\frac{w}{2}})a + e^{\frac{w}{2}}} \right)$$

and

$$A_{[a,b]}(x) = (b-a)x + a.$$

Hence

$$\begin{aligned} Nm|_{[a,b]}(x) &= A_{m[a,b]}^{-1} \circ m \circ A_{[a,b]}(x) \\ &= \frac{[(1-e^{\frac{w}{2}})a+e^{\frac{w}{2}}][(1-e^{\frac{w}{2}})b+e^{\frac{w}{2}}]}{e^{\frac{w}{2}}(b-a)} \left(m((b-a)x+a) - \frac{a}{(1-e^{\frac{w}{2}})a + e^{\frac{w}{2}}} \right) \\ &= \frac{[(1-e^{\frac{w}{2}})a + e^{\frac{w}{2}}][(1-e^{\frac{w}{2}})b + e^{\frac{w}{2}}]}{e^{\frac{w}{2}}(b-a)} \\ &\quad \cdot \frac{e^{\frac{w}{2}}(b-a)x}{[(1-e^{\frac{w}{2}})((b-a)x+a) + e^{\frac{w}{2}}][(1-e^{\frac{w}{2}})a + e^{\frac{w}{2}}]} \\ &= \frac{x}{\frac{(1-e^{\frac{w}{2}})(b-a)}{(1-e^{\frac{w}{2}})b+e^{\frac{w}{2}}} x + \frac{(1-e^{\frac{w}{2}})a+e^{\frac{w}{2}}}{(1-e^{\frac{w}{2}})b+e^{\frac{w}{2}}}} = \frac{x}{\left(1 - \frac{(1-e^{\frac{w}{2}})a+e^{\frac{w}{2}}}{(1-e^{\frac{w}{2}})b+e^{\frac{w}{2}}}\right) x + \frac{(1-e^{\frac{w}{2}})a+e^{\frac{w}{2}}}{(1-e^{\frac{w}{2}})b+e^{\frac{w}{2}}}}. \end{aligned}$$

Namely $Nm|_{[a,b]}(x)$ is a Möbius map of weight w', where w' satisfies

$$e^{\frac{w'}{2}} = \frac{(1-e^{\frac{w}{2}})a + e^{\frac{w}{2}}}{(1-e^{\frac{w}{2}})b + e^{\frac{w}{2}}}.$$

Using (5.1.5) we have

$$e^{w'} = \frac{Dm(b)}{Dm(a)}.$$

Suppose $a = 0$ and $b = \sigma$, we have

$$e^{w'} = \frac{Dm(\sigma)}{Dm(0)} = \frac{e^{w}}{[(1-e^{\frac{w}{2}})\sigma + e^{\frac{w}{2}}]^2} = \frac{1}{[(e^{-\frac{w}{2}} - 1)\sigma + 1]^2}.$$

Hence

$$(e^{-\frac{w'}{2}} - 1) = \sigma(e^{-\frac{w}{2}} - 1).$$

□

We denote by M the set of all Mobius maps of the interval.

Lemma 5.1.6 *The set M together with the composition is a group. Moreover*

$$m_{w_2} \circ m_{w_1} = m_{w_2+w_1}.$$

In particular the function $w : (M, \circ) \to (\mathbb{R}, +)$ defined as $w(m) = \log \frac{Dm(1)}{Dm(0)}$ is a group isomorphism.

The proof is left as an exercise, see Exercise 5.4.2. The definition of Mobius circle maps is as follows.

Definition 5.1.7 A function $f \in \mathcal{D}^2$ is called a Möbius circle diffeomorphism if both branches are Möbius diffeomorphisms, $Nf_-, Nf_+ \in M$. Moreover, to any Möbius circle diffeomorphism f we can then associate 4 numbers $f = (c_f, v_f, w_+, w_-)$ where w_+, w_- are the weight of the Möbius interval maps Nf_+, Nf_- with $w_+ + w_- = 0$. The set of Möbius circle diffeomorphisms is denoted by $\mathcal{M} \subset \mathcal{D}^2$.

Remark 5.1.8 Let $f \in \mathcal{M}$. Since $w_+ + w_- = 0$, the space $\mathcal{M}$ has dimension 3. Note that Lemmas 5.1.5 and 5.1.6 ensure that $\mathcal{M}$ is invariant under renormalization.

Let $N \in \mathbb{N}$, $N > 0$ and let $\underline{\varphi} = \{\varphi_i\}_{i=1}^N$ with $\varphi_i \in \text{Diff}_+^2([0, 1])$. We define the composition $O\underline{\varphi} \in \text{Diff}_+^2([0, 1])$ as

$$O\underline{\varphi} = \varphi_N \circ \cdots \circ \varphi_2 \circ \varphi_1.$$

The partial composition is defined, for all $i \leq j \leq N$, as

$$O_i^j \underline{\varphi} = \varphi_j \circ \cdots \circ \varphi_i.$$

Lemma 5.1.9 *For all $K > 1$ there exists $C > 0$ such that the following holds. For all $N > 0$, if $\underline{\varphi} = \{\varphi_i\}_{i=1}^N$ and $\underline{\psi} = \{\psi_i\}_{i=1}^N$ with $\sum_i \sup_x |\eta_{\varphi_i}(x)| \leq K$ and $\sum_i \sup_x |\eta_{\psi_i}(x)| \leq K$ then*

$$dist_{C^1}(O\underline{\varphi}, O\underline{\psi}) \leq C \sum dist_\eta(\varphi_i, \psi_i).$$

Proof The proof is by induction in N. For $N = 1$ it is enough to use Lemma 5.1.1 which give us a constant C_1 such that

$$dist_{C^1}(\varphi_1, \psi_1) \leq C_1 dist_\eta(\varphi_1, \psi_1). \tag{5.1.8}$$

We define $C_2 = (C_1 K e^K + 1)$. Suppose now that the lemma holds for N with constant C_2. By the mean value theorem, there exists $\theta_0 \in [0, 1]$ such that $DO_2^{N+1}\underline{\psi}(\theta_0) = 1$. Observe also that

$$\begin{aligned}\mathrm{Dist}(O_2^{N+1}\underline{\psi}) &= \max_{x,y}\log\frac{DO_2^{N+1}\underline{\psi}(x)}{DO_2^{N+1}\underline{\psi}(y)}\\ &= \max_{x,y}\left(\log\frac{D\psi_{N+1}(O_2^N\underline{\psi}(x))}{D\psi_{N+1}(O_2^N\underline{\psi}(y))}+\log\frac{D\psi_N(O_2^{N-1}\underline{\psi}(x))}{D\psi_N(O_2^{N-1}\underline{\psi}(y))}\right.\\ &\left.+\cdots+\log\frac{D\psi_2(x)}{D\psi_2(y)}\right)\\ &\le \sum_{i=2}^{N+1}\int_0^1|\eta_{\psi_i}|\le\sum_{i=2}^{N+1}\sup|\eta_{\psi_i}|\le K.\end{aligned}$$

Hence, for every $\theta\in[0,1]$,

$$DO_2^{N+1}\underline{\psi}(\theta)=\frac{DO_2^{N+1}\underline{\psi}(\theta)}{DO_2^{N+1}\underline{\psi}(\theta_0)}\le e^{\mathrm{Dist}(O_2^{N+1}\underline{\psi})}\le e^K. \tag{5.1.9}$$

Consequently if $x\in[0,1]$ and $\Delta=\left|O_1^{N+1}\underline{\varphi}(x)-O_1^{N+1}\underline{\psi}(x)\right|$, then

$$\begin{aligned}\Delta &\le \left|O_2^{N+1}\underline{\varphi}(\varphi_1(x))-O_2^{N+1}\underline{\psi}(\varphi_1(x))\right|+\left|O_2^{N+1}\underline{\psi}(\varphi_1(x))-O_2^{N+1}\underline{\psi}(\psi_1(x))\right|\\ &\le C_2\sum_{k=2}^{N+1} dist_\eta(\varphi_k,\psi_k)+\max_\theta DO_2^{N+1}\underline{\psi}(\theta)\cdot|\varphi_1(x)-\psi_1(x)|\\ &\le C_2\sum_{k=2}^{N+1} dist_\eta(\varphi_k,\psi_k)+e^K\cdot C_1 dist_\eta(\varphi_1,\psi_1)\\ &\le C_2\sum_{k=1}^{N+1} dist_\eta(\varphi_k,\psi_k).\end{aligned}$$

We proved by induction that for all N

$$\sup_x\left|O_1^N\underline{\varphi}(x)-O_1^{N+1}\underline{\psi}(x)\right|\le C_2\sum_{i=1}^N \mathrm{dist}_\eta(\varphi_i,\psi_i).$$

For estimating $\left|DO_1^{N+1}\underline{\varphi}(x)-DO_1^{N+1}\underline{\psi}(x)\right|$ first observe that similar as in (5.1.9) we have that for all $i\le j\le N+1$ and all N,

$$e^{-K}\le|DO_i^j\underline{\varphi}(x)|,\ |DO_i^j\underline{\psi}(x)|\le e^K. \tag{5.1.10}$$

Hence (5.1.8) implies (see Remark 5.1.2) that

$$\left|\log \frac{DO_1^1\underline{\psi}(x)}{DO_1^1\underline{\varphi}(x)}\right| \leq e^K C_1 dist_\eta(\varphi_1, \psi_1) \leq C_2 dist_\eta(\varphi_1, \psi_1).$$

Now assume by induction that for all $\underline{\varphi} = \{\varphi_i\}_{i=1}^N$ and $\underline{\psi} = \{\psi_i\}_{i=1}^N$ with $\sum \sup |\eta_{\varphi_i}| \leq K$ and $\sum \sup |\eta_{\psi_i}| \leq K$ we have

$$\left|\log \frac{DO_1^N\underline{\psi}(x)}{DO_1^N\underline{\varphi}(x)}\right| \leq C_2 \sum_{i=1}^{N} dist_\eta(\varphi_i, \psi_i), \tag{5.1.11}$$

and we prove it for $N+1$. By induction hypothesis we have

$$\begin{aligned}
\left|\log \frac{DO_1^{N+1}\underline{\psi}(x)}{DO_1^{N+1}\underline{\varphi}(x)}\right| &= \left|\log \frac{DO_2^{N+1}\underline{\psi}(\psi_1(x))D\psi_1(x)}{DO_2^{N+1}\underline{\varphi}(\varphi_1(x))D\varphi_1(x)}\right| \\
&\leq \left|\log \frac{DO_2^{N+1}\underline{\psi}(\psi_1(x))}{DO_2^{N+1}\underline{\psi}(\varphi_1(x))}\right| + \left|\log \frac{DO_2^{N+1}\underline{\psi}(\varphi_1(x))}{DO_2^{N+1}\underline{\varphi}(\varphi_1(x))}\right| \\
&\quad + dist_\eta(\varphi_1, \psi_1) \\
&\leq \left|\log \frac{DO_2^{N+1}\underline{\psi}(\psi_1(x))}{DO_2^{N+1}\underline{\psi}(\varphi_1(x))}\right| \\
&\quad + C_2 \sum_{i=2}^{N+1} dist_\eta(\varphi_i, \psi_i) + dist_\eta(\varphi_1, \psi_1)
\end{aligned} \tag{5.1.12}$$

Since

$$\begin{aligned}
\left|\log \frac{DO_2^{N+1}\underline{\psi}(\psi_1(x))}{DO_2^{N+1}\underline{\psi}(\varphi_1(x))}\right| &= \left|\log \frac{D\psi_{N+1}(O_2^N\underline{\psi}(\psi_1(x)))}{D\psi_{N+1}(O_2^N\underline{\psi}(\varphi_1(x)))}\right. \\
&\quad \left. + \log \frac{D\psi_N(O_2^{N-1}\underline{\psi}(\psi_1(x)))}{D\psi_N(O_2^{N-1}\underline{\psi}(\varphi_1(x)))} + \cdots + \log \frac{D\psi_2(\psi_1(x))}{D\psi_2(\varphi_1(x))}\right|,
\end{aligned}$$

let Δ_2 be the interval with endpoints $\varphi_1(x)$, $\psi_1(x)$ and $\Delta_i = O_2^{i-1}\underline{\psi}(\Delta_2)$ for all $i = 3, \ldots, N-1$ then we have

$$\left|\log \frac{DO_2^{N+1}\psi(\psi_1(x))}{DO_2^{N+1}\psi(\varphi_1(x))}\right| \leq \sum_{i=2}^{N} \int_{\Delta_i} |\eta_{\psi_i}| \leq \sum_{i=2}^{N} \sup |\eta_{\psi_i}| \cdot |\Delta_i|.$$

By Eqs. (5.1.10) and (5.1.8) we get

$$|\Delta_i| \le e^K |\Delta_2| \le e^K C_1 \mathrm{dist}_\eta(\varphi_1, \psi_1).$$

Hence

$$\left|\log \frac{DO_2^{N+1}\underline{\psi}(\psi_1(x))}{DO_2^{N+1}\underline{\psi}(\varphi_1(x))}\right| \le K e^K C_1 \mathrm{dist}_\eta(\varphi_1, \psi_1). \tag{5.1.13}$$

By Eqs. (5.1.13) and (5.1.12) we have

$$\begin{aligned}\left|\log \frac{DO_1^{N+1}\underline{\psi}(x)}{DO_1^{N+1}\underline{\varphi}(x)}\right| &\le K e^K C_1 \mathrm{dist}_\eta(\varphi_1, \psi_1) + C_2 \sum_{i=2}^{N+1} \mathrm{dist}_\eta(\varphi_i, \psi_i) + \mathrm{dist}_\eta(\varphi_1, \psi_1)\\ &\le C_2 \sum_{i=1}^{N+1} \mathrm{dist}_\eta(\varphi_i, \psi_i).\end{aligned}$$

This proves (5.1.11). Now again by (5.1.10) and Remark 5.1.2 we have

$$\left|DO_1^N\underline{\varphi}(x) - DO_1^N\underline{\psi}(x)\right| \le e^K \left|\log \frac{DO_1^N\underline{\psi}(x)}{DO_1^N\underline{\varphi}(x)}\right| \le e^K C_2 \sum_{i=1}^{N} dist_\eta(\varphi_i, \psi_i)$$

for all N, and we proved that the lemma holds for all N and the constant $C = e^K C_2$. □

The Möbius diffeomorphisms are good local approximations to diffeomorphisms, better than linear maps, see Fig. 5.1. In particular,

Proposition 5.1.10 *Let $f \in \mathcal{D}^3$. Then there exists a $C > 0$ such that for all intervals $I \subset [0, c_f) \cup (c_f, 1]$*

$$dist_\eta(N(f|_I), m_w) \le C|I|^2$$

where m_w is the Mobius map of weight $w = \int_I \eta_f$.

Proof Let $\varphi = N(f|_I)$. Then using the chain rule for non-linearities (see Lemma 3.2.3) and note that for an affine map A_I, $DA_I(x) \equiv |I|$ and $\eta_{A_I}(x) \equiv 0$, we have

$$\begin{aligned}\eta_\varphi(x) = \eta_{A^{-1}_{f(I)}\circ f\circ A_I}(x) &= \eta_{A^{-1}_{f(I)}\circ f}(A_I(x)) \cdot DA_I(x) + \eta_{A_I}(x)\\ &= \eta_{A^{-1}_{f(I)}\circ f}(A_I(x)) \cdot |I|\\ &= \left(\eta_{A^{-1}_{f(I)}}(f(A_I(x))) \cdot Df(A_I(x)) + \eta_f(A_I(x))\right) \cdot |I|\\ &= \eta_f(A_I(x)) \cdot |I|.\end{aligned}$$

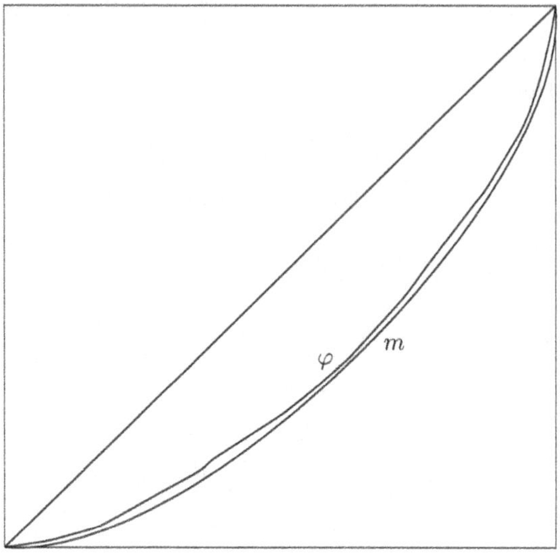

Fig. 5.1 Approximation by Möbius maps

This implies that

$$|D\eta_\varphi(x)| = D\eta_f(A_I(x))|I|^2 \le C|I|^2$$

with the constant $C > 0$ coming from the fact that f is a C^3 diffeomorphism, and

$$\int_0^1 \eta_\varphi(x)dx = \int_0^1 \eta_f(A_I(x)) \cdot DA_I(x)dx = \int_{A_I([0,1])} \eta_f(A_I(x))dA_I(x)$$
$$= \int_I \eta_f(x)dx = w.$$

Applying the mean value theorem to η_φ, there exists $x_0 \in [0, 1]$ such that $\eta_\varphi(x_0) = w$. Thus for any $x \in [0, 1]$,

$$\left|\eta_\varphi(x) - w\right| = |\eta_\varphi(x) - \eta_\varphi(x_0)| \le \frac{|\eta_\varphi(x) - \eta_\varphi(x_0)|}{|x - x_0|} \le \max |D\eta_\varphi(x)| \le C|I|^2. \tag{5.1.14}$$

A similar estimates hold for Möbius maps. Notice that there exists $C_1 > 0$ such that $|w| \le \int_I |\eta_f| \le C_1|I|$. The formula (5.1.7) after calculations gives that there exists a constant $C_2 > 0$ such that

$$\max |D\eta_{m_w}(x)| = 2(e^{\frac{|w|}{2}} - 1)^2 \le C_2 w^2.$$

As before there exists a constant C_3 such that

$$\left|\eta_{m_w}(x) - w\right| \leq C_3|I|^2. \tag{5.1.15}$$

From Eqs. (5.1.14) and (5.1.15),

$$|\eta_\varphi(x) - \eta_{m_w}(x)| \leq (C + C_3)|I|^2 = C_4|I|^2.$$

The lemma follows. □

Remark 5.1.11 Dear reader, as you can see in the previous proof, the use of constants might become quite cumbersome. In such cases one denotes all uniform constants by the same C.

Proposition 5.1.12 *For every $f \in \mathcal{I}^3$ there exists $C > 0$ such that*

$$d_1(R^n f, \mathcal{M}) \leq Cmesh(\mathscr{P}_n) \to 0.$$

More precisely if for all $n \in \mathbb{N}$,

- $c_n = c_{R^n f}$,
- $v_n = v_{R^n f}$,
- $w_{n,+} = \int_{(c_n,1]} \eta_{R^n f}$,
- $w_{n,-} = \int_{[0,c_n)} \eta_{R^n f}$,

let the Mobius map $m_n = (c_n, v_n, w_{n,+}, w_{n,-}) \in \mathcal{M}$ then

$$d_1(R^n f, m_n) \leq Cmesh(\mathscr{P}_n) \to 0.$$

Moreover for all $n \in \mathbb{N}$, $w_{n,-} = \int_{\cup \mathscr{L}_n} \eta_f$, $w_{n,+} = \int_{\cup \mathscr{R}_n} \eta_f$ and

$$w_{n,-} + w_{n,+} = 0.$$

Proof For all $n \in \mathbb{N}$ and for all $i = 0, 1, \ldots, l_n - 1$ let $\varphi_i = Nf|_{f^i(L_n)}$, $w_i = \int_{f^i(L_n)} \eta_f$ and let $m_i \in M$ be the Mobius diffeomorphism with weight w_i. Define $\underline{\varphi} = \{\varphi_i\}_{i=0}^{l_n-1}$ and $\underline{m} = \{m_i\}_{i=0}^{l_n-1}$. By Lemma 5.1.9 and Proposition 5.1.10 we get

$$\begin{aligned} dist_{C^1}(O\underline{\varphi}, O\underline{m}) &\leq C \sum |f^i(L_n)|^2 \leq C(\max |f^i(L_n)) \left|\sum |f^i(L_n)\right| \\ &\leq C\mathrm{mesh}(\mathscr{L}_n). \end{aligned}$$

By Lemma 5.1.6 $O\underline{m}$ is a Möbius diffeomorphism with weight

$$w_{n,-} = \sum w_i = \int_{\cup \mathscr{L}_n} \eta_f.$$

The proposition follows from the fact that $\int_{\cup \mathscr{L}_n} \eta_f = \int_{[0,c_n)} \eta_{R^n f}$. This follows by direct computation. Namely, we have (where we denote by u_n^- the left end point of L_n):

$$\begin{aligned}\int_0^{c_n} \eta_{R^n f} &= \log \frac{DR^n f(c_n)}{DR^n f(0)} = \log \frac{Df^{l_n}(c_f)}{Df^{l_n}(u_n^-)} \\ &= \log \frac{Df(f^{l_n-1}(c_f))}{Df(f^{l_n-1}(u_n^-))} + \log \frac{Df(f^{l_n-2}(c_f))}{Df(f^{l_n-2}(u_n^-))} + \cdots + \log \frac{Df(c_f)}{Df(u_n^-)} \\ &= \sum_{i=0}^{l_n-1} \int_{f^i(L_n)} \eta_f = \int_{\cup \mathscr{L}_n} \eta_f .\end{aligned}$$

The left branch of $R^n f$ can be treated in the same way. □

5.2 From Möbius Maps to Affine Maps

Definition 5.2.1 A function $f \in \mathcal{D}^2$ is called an affine circle map if the normalized branches are the identity function, $Nf_- = \text{id}$ and $Nf_+ = \text{id}$. The set of affine circle maps is denoted by $\mathcal{A} \subset \mathcal{M} \subset \mathcal{D}^2$.

Lemma 5.2.2 *Let $m \in \mathcal{M}$, $m = (c, v, w_+, w_-)$ then*

$$d_\eta(m, \mathcal{A}) = |w_+| = |w_-|.$$

Proof Observe that functions in $\mathcal{A} \subset \mathcal{M}$ are characterized by the property that $w_+ = w_- = 0$. Choose $f \in \mathcal{A}$ such that $c_f = c$, $v_f = v$. Then

$$d_\eta(m, \mathcal{A}) = \max\{dist_\eta(m_{w_+}, \text{id}), dist_\eta(m_{w_-}, \text{id})\} = \max\left\{\int_0^1 |\eta_{m_{w_+}}|, \int_0^1 |\eta_{m_{w_-}}|\right\}.$$

By (5.1.7) we have that $\eta_{m_w} \geq 0$ if and only if $w \geq 0$. Hence, $\int |\eta_{m_w}| = |\int \eta_{m_w}| = |w|$ for all w. Then $d_\eta(m, \mathcal{A}) = \max\{|w_+|, |w_-|\} = |w_+| = |w_-|$. □

Proposition 5.2.3 *Let $m \in \mathcal{I}^2 \cap \mathcal{M}$ then there exists a constant $C > 0$ such that*

$$d_\eta(R^n m, \mathcal{A}) \leq C mesh(\mathscr{P}_n) \longrightarrow 0$$

when $n \to \infty$.

Proof For $n \in \mathbb{N}$ we denote by $w_{n,+}$ the weight of the normalized right branch $m_{n,+}$ of $R^n m$ and by $w_{n,-}$ the weight of the normalized left one $m_{n,-}$. Then by Lemma 5.2.2 $d_\eta(R^n m, \mathcal{A}) = |w_{n,+}| = |w_{n,-}|$. We will prove that

$$0 \leq \frac{w_{n+1,+}}{w_{n,+}} \leq 1. \tag{5.2.1}$$

Observe that by (5.1.6) if $w_{n,+} > 0$ then for all $x \in [0, 1]$, $Dm_{n,+}(0) \leq Dm_{n,+}(x) \leq Dm_{n,+}(1)$ and if $w_{n,+} \leq 0$ then for all $x \in [0, 1]$, $Dm_{n,+}(0) \geq Dm_{n,+}(x) \geq Dm_{n,+}(1)$. Now if $R^n m \in \mathcal{D}^2_-$, observe that $m_{n+1,+} = Nm_{n,+}|_{[0,\sigma_n]}$, where $\sigma_n = \frac{|R_{n+1}|}{|R_n|}$. By (5.1.4) and Lemma 5.1.5 we have

$$0 \leq \frac{w_{n+1,+}}{w_{n,+}} = \frac{\log Dm_{n,+}(\sigma_n) - \log Dm_{n,+}(0)}{\log Dm_{n,+}(1) - \log Dm_{n,+}(0)} \leq 1.$$

Equation (5.2.1) follows. The same argument show that if $R^n m \in \mathcal{D}^2_+$ then

$$0 \leq \frac{w_{n+1,-}}{w_{n,-}} \leq 1. \tag{5.2.2}$$

Since

$$\frac{w_{n+1,+}}{w_{n,+}} = \frac{-w_{n+1,-}}{-w_{n,-}} = \frac{w_{n+1,-}}{w_{n,-}},$$

Equation (5.2.1) follows.

Let $z : \mathbb{R} \to \mathbb{R}$ be the function defined as $z(w) = 1 - e^{-\frac{w}{2}}$. This function z has the following properties:

1. $0 < w' < w \iff 0 < z(w') < z(w)$,
2. $w < w' < 0 \iff z(w) < z(w') < 0$.

Let $z_{n,+} = z(w_{n,+})$ and let $\sigma_n = |R_{n+1}|/|R_n|$. We prove by induction that

$$|z_{n,+}| \leq |z_{0,+}| \prod_{k=0}^{n-1} \sigma_k. \tag{5.2.3}$$

For $n = 0$ this is trivially true. We suppose that it holds for n and we prove it for $n + 1$. We have to consider two cases.

- If $R^n m \in \mathcal{D}^2_-$ then by Lemma 5.1.5 we get

$$z_{n+1,+} = \sigma_n z_{n,+} \tag{5.2.4}$$

  which finishes the induction step.
- If $R^n m \in \mathcal{D}^2_+$ then according to Eq. (5.2.1) $|w_{n+1,+}| \leq |w_{n,+}|$. By properties 1 and 2 of the function z we get $|z_{n+1,+}| \leq |z_{n,+}|$ which concludes the proof because in this case $\sigma_n = 1$.

Let us rewrite Eq. (5.2.3) and by Lemma 4.2.6 we get:

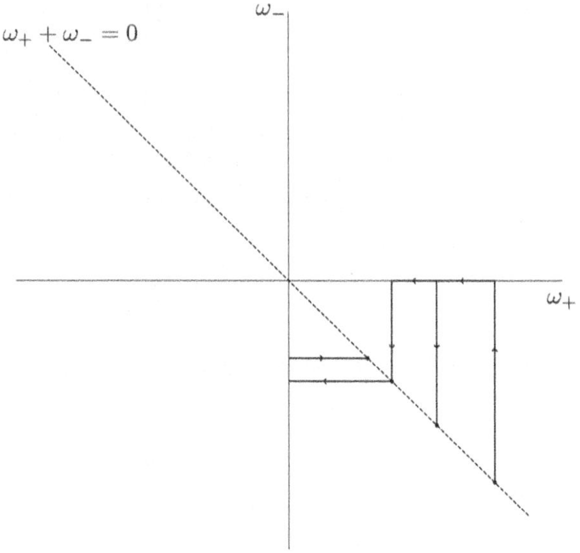

Fig. 5.2 Renormalization of Möbius maps

$$|z_{n,+}| \leq |R_n| \frac{|z_{0,+}|}{|R_0|} \leq C\text{mesh}(\mathscr{P}_n) \longrightarrow 0.$$

Hence, there exists a positive constant $C > 0$ such that

$$|w_{n,-}| = |w_{n,+}| \leq C\text{mesh}(\mathscr{P}_n) \longrightarrow 0.$$

□

Remark 5.2.4 The essence of the explanation for the convergence described in Proposition 5.2.3 is (5.2.4) and $w_+ + w_- = 0$. The renormalization steps are illustrated in Fig. 5.2.

5.3 From Affine Maps to Rotations

Recall that the set of rotations is denoted by $\mathcal{W} \subset \mathcal{D}^2$. The following proposition states the convergence of renormalization of affine maps. Observe that renormalization of affine maps is in essence a piecewise linear map, see (5.3.2) and (5.3.3). The convergence is illustrated in Fig. 5.3.

Proposition 5.3.1 *Let $a \in \mathcal{I}^2 \cap \mathcal{A}$ then*

$$d_\eta(R^n a, \mathcal{W}) \longrightarrow 0$$

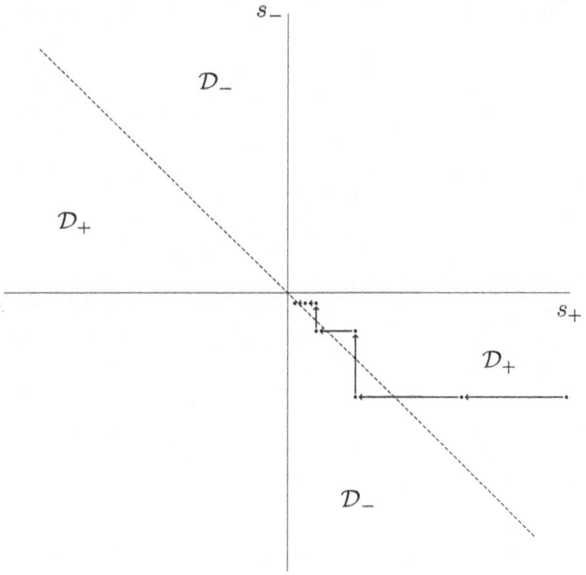

Fig. 5.3 Renormalization of affine maps

monotonically when $n \to \infty$.

Proof For $n \in \mathbb{N}$ we denote by $a_{n,+}$ and $a_{n,-}$ the right and left branches of $R^n a$ and by

$$s_{n,+} = \log Da_{n,+}, \quad s_{n,-} = \log Da_{n,-}.$$

Note that it suffices to prove that $|s_{n,-}| \to 0$ monotonically:

$$\begin{aligned} d_\eta(R^n a, \mathcal{W}) &= \min_{r_\rho \in \mathcal{W}} d_\eta(R^n a, r_\rho) \\ &= \min_{\rho \in [0,1]} (\max\{|c_{R^n a} - (1-\rho)|, |v_{R^n a} - \rho|, 0, 0\}) \\ &\le |c_{R^n a} - (1 - v_{R^n a})| \\ &\le c_{R^n a}|(1 - Da_{n,-})|, \end{aligned}$$

which tends to 0 when $s_{n,-} \to 0$.

Observe that the two branches cannot both have derivatives bigger or both smaller than 1, namely

$$s_{n,+} s_{n,-} \le 0. \tag{5.3.1}$$

The definition of renormalization implies immediately that

$$R^n a \in \mathcal{D}_+^2 \implies \begin{cases} s_{n+1,+} = s_{n,+} + s_{n,-} \\ s_{n+1,-} = s_{n,-} \end{cases} \tag{5.3.2}$$

$$R^n a \in \mathcal{D}_-^2 \implies \begin{cases} s_{n+1,+} = s_{n,+} \\ s_{n+1,-} = s_{n,+} + s_{n,-} \end{cases} \tag{5.3.3}$$

Equations (5.3.1), (5.3.2), and (5.3.3) and the fact that $a_n(a) < \infty$ for all n implies that $|s_{n,\pm}|$ converges monotonically to a limit which has to be a fixed point of Eqs. (5.3.2) and (5.3.3). Hence $|s_{n,-}| \longrightarrow 0$, compare Fig. 5.3. This proves the proposition. □

5.4 Exercises

Exercise 5.4.1 Prove Lemma 5.1.4.

Exercise 5.4.2 Prove Lemma 5.1.6.

Exercise 5.4.3 Determine the set of piece wise affine circle maps of type

$$\sigma = (+, +, -, +, +, -, \cdots).$$

Exercise 5.4.4 Give a proof of the exponential convergence of renormalization to rotations of piece wise affine circle maps with periodic renormalization type (the sequence σ is periodic).

Exercise 5.4.5 Show that in the space of affine circle maps the maps with the same rotation number form a vector space.

Chapter 6
The Collaboration of the Three Ingredients

In this chapter we prepare the ground to prove that consecutive renormalizations of circle diffeomorphisms converge to the set of rotations. In the previous chapter we described the three ingredients which are responsible for this convergence, namely circle diffeomophisms converge to the set of Möbius circle maps, Möbius circle maps converge to the set of affine circle maps and affine circle maps converge to the set of rotations. Now it is time to merge these three ingredients and show their collaboration. Indeed, these mechanisms are also active when one renormalizes generalized diffeomorphisms. Different from the specific cases of Möbius maps and affine maps where the effects of the ingredients are described by exact formulas, when applying them in the general case of circle diffeomorphisms one gets errors. This is described in Proposition 6.1.3 where one can clearly recognize the ingredients responsible for convergence in the cases of affine maps and the Möbius maps. Fortunately, the errors appearing in the generalized context, decay very fast.

The main results in this chapter are Proposition 6.1.3 and the decay of the error terms which is established in Propositions 6.2.7 and 6.2.8.

6.1 The Collaboration

We start introducing some notation which is inspired by the previous chapter but is now applied to the general case of diffeomorphisms. We fix $f \in \mathcal{I}^3$ and we define the right slope of f as $d_+ = \frac{v}{1-c}$, the left one as $d_- = \frac{1-v}{c}$ and their logarithms as

$$s_\pm = \log d_\pm .$$

Y. Dong et al., *One-Dimensional Dynamics*, Lecture Notes of the Unione Matematica Italiana 30, https://doi.org/10.1007/978-3-031-94882-4_6

Let $w = \int_c^1 \eta_f$ and let

$$z = e^{-\frac{w}{2}} - 1.$$

We will constantly use the scaling ratio σ defined as

$$\sigma = \begin{cases} \frac{v-c}{1-c} & \text{if } f \in \mathcal{D}_-^3 \\ \frac{c-v}{c} & \text{if } f \in \mathcal{D}_+^3 \end{cases}$$

In the same spirit we define for all $n \in \mathbb{N}$ the slopes of the nth renormalization $R^n f$, $d_+(n)$ and $d_-(n)$. Similarly for $R^n f$ we define $s_\pm(n)$, $w(n)$, $z(n)$ and $\sigma(n)$. Finally let $\delta(n) = \text{mesh}(\mathscr{P}_n)$, $\delta_+(n) = \text{mesh}(\mathscr{R}_n)$ and $\delta_-(n) = \text{mesh}(\mathscr{L}_n)$. By Lemma 4.2.6 we have

$$\delta(n) \to 0. \tag{6.1.1}$$

By Lemma 4.6.1 we have for some constant $C > 0$

$$|w(n)| = \int_{c_n}^1 \eta_{R^n f} = \log \frac{DR^n f(1)}{DR^n f(c_n)} \leq \text{Dist}(R^n f) \leq C.$$

Hence by the mean value theorem (recall Remark 5.1.2) we have the following bounds:

$$\frac{1}{K} \leq \frac{|z(n)|}{|w(n)|} \leq K \tag{6.1.2}$$

and

$$|z(n)|, |w(n)| \leq K, \tag{6.1.3}$$

where K is a uniform constant which depends on f and doesn't depend on n.

Lemma 6.1.1 *Let $f \in \mathcal{I}^3$, then*

$$\sigma = \begin{cases} \frac{1-d_-d_+}{1-d_-} & \text{if } f \in \mathcal{D}_-^3 \\ \frac{1-d_-d_+}{1-d_+} & \text{if } f \in \mathcal{D}_+^3 \end{cases}$$

Proof Let $f \in \mathcal{D}_-^3$. If we denote $T = 1 - c$ and $M = v - c$ then by definition $\sigma = \frac{M}{T}$. Observe that $Td_+ + (1 - T)d_- = 1$. Hence

$$T = \frac{1 - d_-}{d_+ - d_-}.$$

Moreover

$$M = T - d_-(1 - T) = T(1 + d_-) - d_-$$

and consequently

$$\sigma = \frac{M}{T} = \frac{1 - d_- d_+}{1 - d_-}.$$

The case $f \in \mathcal{D}_+^3$ can be treated similarly. □

The following lemma clarifies how the space $\mathcal{I}^3$ and the subspaces $\mathcal{D}_-^3, \mathcal{D}_+^3$ project onto the coordinate plane spanned by s_+, s_-, see Fig. 5.3.

Lemma 6.1.2 *Let $f \in \mathcal{I}^3$, then*

$$s_- s_+ < 0.$$

Moreover

$$f \in \mathcal{D}_-^3 \implies \begin{cases} 0 < s_+ < -s_- & if\, s_- < 0 \\ 0 < -s_+ < s_- & if\, s_- > 0 \end{cases}$$

$$f \in \mathcal{D}_+^3 \implies \begin{cases} s_+ > -s_- > 0 & if\, s_- < 0 \\ -s_+ > s_- > 0 & if\, s_- > 0 \end{cases}$$

Proof Observe that

$$d_+(1 - c) + cd_- = 1.$$

This implies immediately that d_+ and d_- cannot be both larger than 1 neither both smaller than 1. Hence $s_- s_+ < 0$. Let $f \in \mathcal{D}_-^3$ and $s_- < 0$. Consequently $d_- < 1$. By Lemma 6.1.1 $\sigma = \frac{1-d_- d_+}{1-d_-}$, and observe that by definition $\sigma > 0$, we have

$$1 - d_- d_+ > 0$$

and

$$s_- + s_+ < 0.$$

The other cases can be treated similarly. □

As we mentioned in the introduction, the behaviour of the non-linearity $z(n)$ of $R^n f$ differs from the behaviour of the renormalizations of Möbius maps by an error term, and the behaviour of the slopes $s_\pm(n)$ of $R^n f$ differs from the behaviour of the renormalizations of affine maps by an error term. The precise statement is described in the following proposition. Compare with Proposition 5.2.3, in particular (5.2.4), which is the second ingredient. The third ingredient, see Proposition 5.3.1, in particular (5.3.2) and (5.3.3) can also be recognized in the proposition below.

Proposition 6.1.3 *Let $f \in \mathcal{I}^3$. There exists a constant $C > 0$ such that the following holds:*

$$|z(n+1) - \sigma(n)z(n))| \le C\delta(n).$$

Moreover, if $R^n f \in \mathcal{D}_-^3$ then

$$|s_+(n+1) - s_+(n)| \le C(|z(n)| + \delta(n))$$
$$|s_-(n+1) - (s_+(n) + s_-(n))| \le C(|z(n)| + \delta(n)).$$

If $R^n f \in \mathcal{D}_+^3$ then

$$|s_+(n+1) - (s_+(n) + s_-(n))| \le C(|z(n)| + \delta(n))$$
$$|s_-(n+1) - s_-(n)| \le C(|z(n)| + \delta(n)).$$

Proof Let $R^n f \in \mathcal{D}_-^3$, $c_n = c_{R^n f}$ and $v_n = v_{R^n f}$. Then

$$d_+(n) = \frac{v_n}{1 - c_n}.$$

Consider $\varphi = N(R^n f|_{[c_n,1]}) : [0, 1] \to [0, 1]$ then $\int \eta_\varphi = w(n)$. Let $m : [0, 1] \to [0, 1]$ be the Möbius map with weight $\int \eta_\varphi = w(n)$. From Proposition 5.1.12 we have

$$dist_{\mathcal{C}^1}(\varphi, m) \le C\delta(n).$$

Hence, for all $x \in [0, 1]$

$$\left|\log \frac{D\varphi(x)}{Dm(x)}\right| \le C|D\varphi(x) - Dm(x)| \le C\delta(n). \tag{6.1.4}$$

Let $\omega = \log \frac{Dm(\sigma(n))}{Dm(0)}$. Then by Lemma 5.1.5,

$$(e^{-\frac{\omega}{2}} - 1) = \sigma(n)(e^{-\frac{w(n)}{2}} - 1) = \sigma(n)z(n).$$

By definition

$$w(n+1) = \log \frac{D\varphi(\sigma(n))}{D\varphi(0)} = \log \frac{D\varphi(\sigma(n))}{Dm(\sigma(n))} + \log \frac{Dm(0)}{D\varphi(0)} + \omega.$$

Hence, by (6.1.4)

$$|w(n+1) - \omega| \leq C\delta(n).$$

We have

$$\begin{aligned}
|z(n+1) - \sigma(n)z(n)| &= |(e^{-\frac{w(n+1)}{2}} - 1) - (e^{-\frac{\omega}{2}} - 1)| \\
&= |e^{-\frac{w(n+1)}{2}} - e^{-\frac{\omega}{2}}| \\
&\leq C|w(n+1) - \omega| \\
&\leq C\delta(n),
\end{aligned}$$

where we used (6.1.3).

By the mean value theorem there exists $x_+ \in [c_n, 1]$ and corresponding $\theta_+ \in [0, \sigma(n)]$ such that

$$d_+(n+1) = DR^n f(x_+) = D\varphi(\theta_+)\frac{v_n}{1-c_n} = D\varphi(\theta_+)d_+(n),$$

and there exists $\theta_0 \in [0, 1]$ such that $Dm(\theta_0) = 1$. Consequently

$$\begin{aligned}
|s_+(n+1) - s_+(n)| &= |\log D\varphi(\theta_+)| \\
&= \left|\log \frac{D\varphi(\theta_+)}{Dm(\theta_+)} + \log \frac{Dm(\theta_+)}{Dm(\theta_0)}\right| \\
&\leq C\delta(n) + |w(n)| \\
&\leq C(\delta(n) + |z(n)|)
\end{aligned}$$

where we used (6.1.2). The first equation for the slopes is then proved.

By the mean value theorem there exists $\theta_- \in [\sigma(n), 1]$ such that

$$d_-(n+1) = d_-(n)D\varphi(\theta_-)d_+(n).$$

Consequently, as before

$$|s_-(n+1) - (s_+(n) + s_-(n))| = |\log D\varphi(\theta_-)| \leq C(\delta(n) + |z(n)|).$$

□

6.2 Fast Renormalization and Error Terms

In general, the convergence of renormalization is not exponentially fast. However, there is a subsequence of remormalizations, topologically defined by the renormalization type of the map, which is exponentially converging. These renormalizations are called the fast renormalizations. The definition of fast renormalization of a map

$f \in \mathcal{I}^3$ with renormalization type $a(f) = \{a_n(f)\}_{n\in\mathbb{N}}$ is the following: if $f \in \mathcal{D}^3_-$, then $fastRf = R^a f$ where a is minimal such that $R^a f \in \mathcal{D}^3_+$ and if $f \in \mathcal{D}^3_+$, then $fastRf = R^a f$ where a is minimal such that $R^a f \in \mathcal{D}^3_-$. In the general case the nth fast renormalization of f is

$$fastR^n f = R^{\alpha_n(f)} f,$$

where the quantity

$$\alpha_n(f) = \begin{cases} \sum_{k=1}^{n+1} a_k(f) & \text{if } f \in \mathcal{D}^3_- \\ \sum_{k=1}^{n} a_k(f) & \text{if } f \in \mathcal{D}^3_+ \end{cases}$$

is called the nth fast renormalizations level. To simplify the notation, in future discussions we will write a_n and α_n instead of $a_n(f)$ and $\alpha_n(f)$ when the function f is clear from the context. We first introduce some more notations.

For a fixed map $f \in \mathcal{I}^3$, we denote by u_n^- and by u_n^+ respectively the left hand point of L_{α_n} and the right hand point of R_{α_n}. Moreover,

- if $fastR^n f \in \mathcal{D}^3_-$ then for all $0 \le i \le a_{n+1}$ we define $u_{n,i}^+ = (f^{r_{\alpha_n}})^i(u_n^+)$
- if $fastR^n f \in \mathcal{D}^3_+$ then for all $0 \le i \le a_{n+1}$ we define $u_{n,i}^- = (f^{l_{\alpha_n}})^i(u_n^-)$.

These points are illustrated in Fig. 6.1.

As an intermezzo, we will briefly discuss how fast-renormalization changes the tower representation of $\mathscr{P}_{\alpha_n}$ into $\mathscr{P}_{\alpha_{n+1}}$. We assume that $fastR^n f \in \mathcal{D}^3_-$. The other case is similar. For $1 \le i \le a_{n+1} + 1$ we define

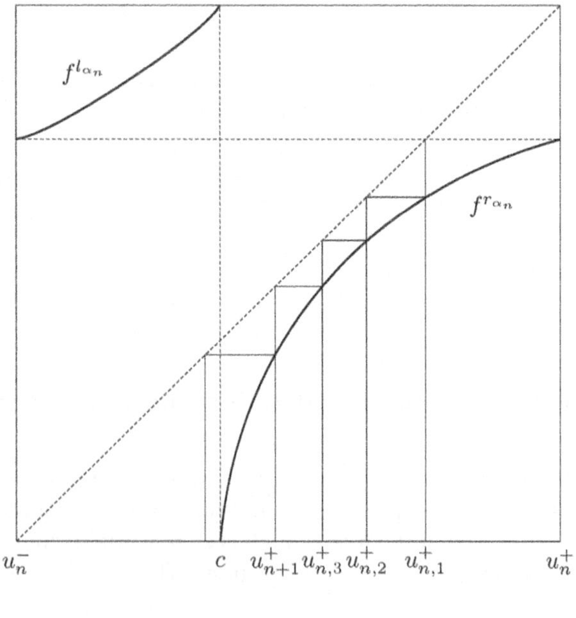

Fig. 6.1 The points $u_n^\pm$ and $u_{n,i}^+$

$$U_{n,i} := \begin{cases} [u^+_{n,i}, u^+_{n,i-1}] & 1 \le i \le a_{n+1}, \\ [c, u^+_{n+1}] & i = a_{n+1} + 1, \end{cases}$$

and $U_{n,i,k} := f^k(U_{n,i})$ for $k < r_{\alpha_n}$.

Remark 6.2.1 Observe that $\mathscr{P}_{\alpha_{n+1}}$ is a refinement of $\mathscr{P}_{\alpha_n}$. In particular,

$$\mathscr{R}_{\alpha_{n+1}} = \left\{U_{n,a_{n+1}+1,k} | 0 \le k < r_{\alpha_n}\right\}$$

and

$$\mathscr{L}_{\alpha_{n+1}} = \mathscr{L}_{\alpha_n} \cup \bigcup_{1 \le i \le a_{n+1}} \left\{U_{n,i,k} | 0 \le k < r_{\alpha_n}\right\}.$$

Fast-renormalization cuts off the pieces $\left\{U_{n,i,k} | 0 \le k < r_{\alpha_n}\right\}$ of the right tower and stacks them on top of the left tower to obtain the tower representation of $\mathscr{P}_{\alpha_{n+1}}$, see Fig. 6.2.

Suppose $fastR^n f \in \mathcal{D}^3_-$ then for all $0 \le i \le a_{n+1}$ the right branch of $R^{\alpha_n + i} f$ is (up to rescaling) a restriction of the right branch of $R^{\alpha_n} f$, which is exactly the same situation as in "slow" renormalization. The case $fastR^n f \in \mathcal{D}^3_+$ is analogous. In Proposition 6.1.3 we see that the non-linearity and the slopes are controlled by the previous term with an error term. Applying the same argument we will show in the following two lemmas that the behaviour of the intermediate levels are controlled by its nearby fast renormalization level with the same error.

Lemma 6.2.2 *Let $f \in \mathcal{I}^3$. There exists a constant $C > 0$ such that for all $0 \le i \le a_{n+1}$,*

$$|z(\alpha_n + i)| \le \sigma_{n,i} |z(\alpha_n)| + C\delta(\alpha_n),$$

where

$$\sigma_{n,i} = \begin{cases} \frac{u^+_{n,i} - c}{u^+_n - c} & R^{\alpha_n} f \in \mathcal{D}^3_- \\ \frac{c - u^-_{n,i}}{c - u^-_n} & R^{\alpha_n} f \in \mathcal{D}^3_+ \end{cases}.$$

Proof We discuss the case $R^{\alpha_n} f \in \mathcal{D}^3_-$. The argument for the other case is similar. Let $\varphi = N(R^{\alpha_n} f|_{(c_{\alpha_n}, 1]})$ and let m be the Möbius map of weight $w(\alpha_n)$. Note that

$$w(\alpha_n + i) = \log \frac{D\varphi(\sigma_{n,i})}{D\varphi(0)} = \log \frac{D\varphi(\sigma_{n,i})}{Dm(\sigma_{n,i})} + \log \frac{Dm(0)}{D\varphi(0)} + \log \frac{Dm(\sigma_{n,i})}{Dm(0)}.$$

Write $\omega_i = \log \frac{Dm(\sigma_{n,i})}{Dm(0)}$, then by Proposition 5.1.12 (see (6.1.4)) we have $|w(\alpha_n + i) - \omega_i| \le C\delta(\alpha_n)$. By Lemma 5.1.5 we have

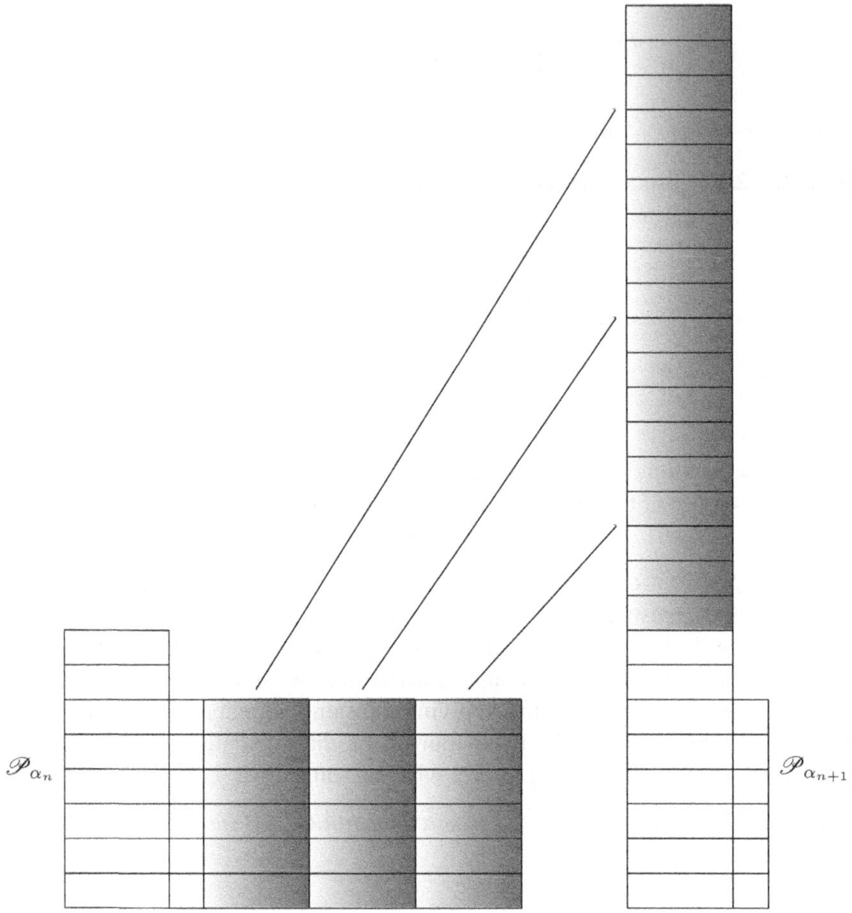

Fig. 6.2 Fast renormalization of towers

$$e^{-\frac{\omega_i}{2}} - 1 = \sigma_{n,i}(e^{-\frac{w(\alpha_n)}{2}} - 1) = \sigma_{n,i} z(\alpha_n).$$

Hence

$$\begin{aligned} |z(\alpha_n + i)| &= \left| e^{-\frac{w(\alpha_n+i)}{2}} - 1 \right| = \left| e^{-\frac{w(\alpha_n+i)}{2}} - e^{-\frac{\omega_i}{2}} + e^{\frac{\omega_i}{2}} - 1 \right| \\ &\leq C|w(\alpha_n + i) - \omega_i| + \sigma_{n,i}|z(\alpha_n)| \\ &\leq C\delta(\alpha_n) + \sigma_{n,i}|z(\alpha_n)|. \end{aligned}$$

□

Lemma 6.2.3 *Let $f \in \mathcal{I}^3$. There exists a constant $C > 0$ with the following properties: For every $n \in \mathbb{N}$, if $R^{\alpha_n} f \in \mathcal{D}^3_-$ then for all $0 \leq i, j \leq a_{n+1}(f)$,*

$$|s_+(\alpha_n + j) - s_+(\alpha_n + i)| \le C(|z(\alpha_n)| + \delta(\alpha_n))$$

and if $R^{\alpha_n} f \in \mathcal{D}^3_+$ *then for all* $0 \le i, j \le a_{n+1}(f)$,

$$|s_-(\alpha_n + j) - s_-(\alpha_n + i)| \le C(|z(\alpha_n)| + \delta(\alpha_n)).$$

Proof Suppose $R^{\alpha_n} f \in \mathcal{D}^3_-$. Let $\varphi = N(R^{\alpha_n} f|_{[c_{\alpha_n}, 1]})$ and m be the Möbius map of weight $w(\alpha_n)$. By the mean value theorem for all $0 \le j \le a_{n+1}(f)$ there exists $x_{n,j} \in [c_{\alpha_n}, 1]$ and corresponding $\theta_{n,j} \in [0, 1]$ such that

$$d_+(\alpha_n + j) = DR^{\alpha_n}(x_{n,j}) = D\varphi(\theta_{n,j}) d_+(\alpha_n).$$

Hence

$$\begin{aligned} |s_+(\alpha_n + j) - s_+(\alpha_n)| &= \left|\log D\varphi(\theta_{n,j})\right| \\ &= \left|\log \frac{D\varphi(\theta_{n,j})}{Dm(\theta_{n,j})} + \log Dm(\theta_{n,j})\right| \\ &\le C\delta(\alpha_n) + C|z(\alpha_n)|, \end{aligned}$$

and for all $0 \le i, j \le a_{n+1}(f)$,

$$\begin{aligned} |s_+(\alpha_n + j) - s_+(\alpha_n + i)| &\le |s_+(\alpha_n + j) - s_+(\alpha_n)| + |s_+(\alpha_n + i) - s_+(\alpha_n)| \\ &\le C(|z(\alpha_n)| + \delta(\alpha_n)). \end{aligned}$$

The case $R^{\alpha_n} f \in \mathcal{D}^3_+$ is analogous.

□

In the rest of this section we will prove that the error terms in Proposition 6.1.3 converge to 0 as n goes to infinity, and the error terms for the fast renormalizations (in Lemmas 6.2.2 and 6.2.3) converge exponentially. More precisely,

$$\delta(n) \to 0, |z(n)| \to 0,$$

and

$$\delta(\alpha_n) \to 0, |z(\alpha_n)| \to 0$$

exponentially.

Recall that $\delta(n) \to 0$. Since the dynamical partition $\mathscr{P}_n$ consists of the iterates of L_n, R_n under f, to prove $\delta(\alpha_n) = \text{mesh}(\mathscr{P}_{\alpha_n}) \to 0$ exponentially fast we naturally would like to obtain a bound for the derivative of fast renormalizations. Recall that the renormalization of $f \in \mathcal{I}^3$ has bounded distortion, see Lemma 4.6.1. In particular, the derivatives of the normalized branches $\varphi_\pm$ of $R^n f$ are always bounded. However, this does not imply that the derivative of $R^n f$ is bounded, as the slopes $d_\pm(n)$ can be very close to 0 or ∞. Nevertheless we will prove in Proposition 6.2.5 that the derivatives of fast renormalizations of f are indeed bounded.

Let C_0 be a constant such that Proposition 6.1.3 and Lemma 6.2.3 hold and let

$$K(n) = C_0(|z(n)| + \delta(n)).$$

For all $n \in \mathbb{N}$, we define $s(n) = (s_+(n), s_-(n))$, and let $|s(n)|$ be the L^1 norm $|s_+(n)| + |s_-(n)|$. To prove that fast renormalization has bounded derivatives we need the following lemma:

Lemma 6.2.4 *Let $f \in \mathcal{I}^3$. If $R^n f \in \mathcal{D}^3_-$ and $|s_+(n)| > 2K(n)$, or if $R^n f \in \mathcal{D}^3_+$ and $|s_-(n)| > 2K(n)$ then*

$$|s(n+1)| < |s(n)|.$$

Proof Suppose $R^n f \in \mathcal{D}^3_-$ and $s_+(n) > 0$. In particular, $s_+(n) > 2K(n)$ and $s_-(n) < 0$. By Proposition 6.1.3 we have

$$s_+(n) - K(n) \le s_+(n+1) \le s_+(n) + K(n)$$
$$s_+(n) + s_-(n) - K(n) \le s_-(n+1) \le s_+(n) + s_-(n) + K(n).$$

From the first inequality we have $s_+(n+1) > 0$, then $s_-(n+1) < 0$. Hence

$$\begin{aligned} |s_+(n+1) + |s_-(n+1)| &= s_+(n+1) - s_-(n+1) \\ &\le s_+(n) + K(n) - (s_+(n) + s_-(n) - K(n)) \\ &= s_+(n) - s_-(n) + (2K(n) - s_+(n)) \\ &\le |s_+(n)| + |s_-(n)|. \end{aligned}$$

The other cases are treated similarly. □

A central theme in renormalization theory are the so-called a priori bounds. In general, one would like that the consecutive renormalizations stay in a compact set. Then at least one can take limits. Indeed, in our present context these a priori bounds have the following form. Namely,

Proposition 6.2.5 *Let $f \in \mathcal{I}^3$, then there exists a constant $K_f > 0$ such that*

$$\frac{1}{K_f} \le DfastR^n f \le K_f.$$

Proof It suffices to show that $|s(\alpha_n)|$ is bounded for all n. In fact, by the mean value theorem there exists $\theta_n \in [c_n, 1]$ such that $\log DR^{\alpha_n} f(\theta_n) = s_+(\alpha_n)$. By

Lemma 4.6.1 the distortion of the renormalization is bounded by a constant C, if $|s(\alpha_n)|$ is bounded then we have

$$|\log DR^{\alpha_n} f(x)| \leq |s_+(\alpha_n)| + \mathrm{Dist}(R^{\alpha_n} f) \leq |s(\alpha_n)| + C$$

is bounded.

From (6.1.1) and (6.1.3) we have that $K(n)$ is bounded. Let $K = \sup_n K(n)$.

Suppose $R^{\alpha_n} f \in \mathcal{D}^3_-$. If $|s_+(\alpha_n+i)| > 2K$ for all $0 \leq i \leq a_{n+1}$, by Lemma 6.2.4 we have

$$|s(\alpha_{n+1})| < |s(\alpha_n + a_{n+1} - 1)| < \cdots < |s(\alpha_n)|.$$

If there exists $0 \leq i \leq a_{n+1}$ such that $|s_+(\alpha_n + i)| \leq 2K$, then by Lemma 6.2.3 we have that

$$|s_+(\alpha_{n+1}) - s_+(\alpha_n + i)| \leq K,$$

hence

$$-3K \leq s_+(\alpha_{n+1}) \leq 3K.$$

Notice that $R^{\alpha_{n+1}} f \in \mathcal{D}^3_+$, by Lemma 6.1.2 we have

$$|s_-(\alpha_{n+1})| < |s_+(\alpha_{n+1})|,$$

hence in this case $|s(\alpha_{n+1})| < 2|s_+(\alpha_{n+1})| \leq 6K$.

Applying the similar argument for the case $R^{\alpha_n} f \in \mathcal{D}^3_+$ and we have proved that

$$|s(\alpha_{n+1})| \leq \max\{|s(\alpha_n)|, 6K\},$$

and therefore

$$|s(\alpha_n)| \leq \max\{|s_-| + |s_+|, 6K\}$$

is bounded for all n. □

Lemma 6.2.6 *Let $f \in \mathcal{I}^3$. There exists $\sigma < 1$ such that for all $n \in \mathbb{N}$,*

- *if fast$R^n f \in \mathcal{D}^3_-$ then $\frac{|u^+_{n+1} - c|}{|u^+_n - c|} \leq \sigma$,*
- *if fast$R^n f \in \mathcal{D}^3_+$ then $\frac{|u^-_{n+1} - c|}{|u^-_n - c|} \leq \sigma$.*

Moreover, if $a_{n+1} \geq 2$ then for all $1 \leq i \leq a_{n+1}$,

- *if fast$R^n f \in \mathcal{D}^3_-$ then $\frac{|u^+_{n,i} - u^+_{n,i-1}|}{|u^+_n - c|} \leq \sigma$,*
- *if fast$R^n f \in \mathcal{D}^3_+$ then $\frac{|u^-_{n,i} - u^-_{n,i-1}|}{|u^-_n - c|} \leq \sigma$.*

And, if $a_{n+1} = a_{n+2} = 1$ *then*

- *if* $fastR^n f \in \mathcal{D}^3_-$ *then* $\frac{|u^+_{n+1} - u^+_n|}{|u^+_n - c|} \le \sigma$,
- *if* $fastR^n f \in \mathcal{D}^3_+$ *then* $\frac{|u^-_{n+1} - u^-_n|}{|u^-_n - c|} \le \sigma$.

Proof From Proposition 6.2.5 we know that there exists constant $K > 0$ such that $\frac{1}{K} \le DfastR^n f \le K$. Observe, in the case $fastR^n f \in \mathcal{D}^3_-$,

$$f^{r_{\alpha_n}}(u^+_{n+1}) < c < u^+_{n+1} \le f^{r_{\alpha_n}}(u^+_n) < u^+_n.$$

Thus we have

$$|f^{r_{\alpha_n}}(u^+_{n+1}) - u^+_n| > |u^+_n - c|,$$

and

$$|f^{r_{\alpha_n}}(u^+_{n+1}) - u^+_n| \le |f^{r_{\alpha_n}}(u^+_{n+1}) - f^{r_{\alpha_n}}(u^+_n)| + |u^+_{n+1} - u^+_n| \le (1+K)|u^+_{n+1} - u^+_n|.$$

Hence,

$$(1 + K)|u^+_{n+1} - u^+_n| > |u^+_n - c|,$$

which implies the first statement of the lemma. The second follows similarly.

Suppose $a_{n+1} \ge 2$ and $fastR^n f \in \mathcal{D}^3_-$. Let $1 < i \le a_{n+1}$. Then

$$|u^+_{n,i} - u^+_{n,i-1}| + |u^+_{n,i-1} - u^+_{n,i-2}| < |u^+_n - c|.$$

By Proposition 6.2.5 we have

$$|u^+_{n,i-1} - u^+_{n,i-2}| \ge \frac{1}{K}|u^+_{n,i} - u^+_{n,i-1}|,$$

hence

$$(1 + \frac{1}{K})|u^+_{n,i} - u^+_{n,i-1}| < |u^+_n - c|.$$

The fourth statement of the lemma follows similarly.

Suppose $a_{n+1} = a_{n+2} = 1$ and $fastR^n f \in \mathcal{D}^3_-$. Then $fastR^{n+1} f \in \mathcal{D}^3_+$. By the second statement of this lemma there exists $\sigma < 1$ such that

$$\frac{|u^-_{n+2} - c|}{|u^-_{n+1} - c|} = \frac{|u^-_{n+2} - c|}{|u^-_{n+1} - u^-_{n+2}| + |u^-_{n+2} - c|} = \frac{1}{\frac{|u^-_{n+1} - u^-_{n+2}|}{|u^-_{n+2} - c|} + 1} \le \sigma,$$

hence

$$\frac{|u_{n+1}^{-}-u_{n+2}^{-}|}{|u_{n+2}^{-}-c|}\geq\frac{1}{\sigma}-1>0. \tag{6.2.1}$$

Suppose $|u_n^+ - u_{n+1}^+| = b|u_n^+ - c|$ for some $0 < b < 1$. We will show that (6.2.1) cannot hold if b is too close to 1. In fact, from $a_{n+1} = a_{n+2} = 1$ we have $f^{r_{\alpha_n}}(c) = u_{n+1}^-$, $f^{r_{\alpha_n}}(u_n^+) = u_{n+1}^+$, and $f^{r_{\alpha_n}}(u_{n+1}^+) = u_{n+2}^-$. By Proposition 6.2.5 we have estimates

$$|u_{n+1}^{-}-u_{n+2}^{-}|\leq K|u_{n+1}^{+}-c|, \tag{6.2.2}$$

and

$$|u_{n+1}^{+}-u_{n+2}^{-}|\geq\frac{1}{K}|u_{n}^{+}-u_{n+1}^{+}|.$$

Notice that

$$|u_{n}^{+}-u_{n+1}^{+}|=\frac{b}{1-b}|u_{n+1}^{+}-c|,$$

and thus

$$|u_{n+2}^{-}-c|=|u_{n+1}^{+}-u_{n+2}^{-}|-|u_{n+1}^{+}-c|\geq(\frac{b}{K(1-b)}-1)|u_{n+1}^{+}-c|. \tag{6.2.3}$$

By inequalities (6.2.1), (6.2.2), and (6.2.3) we have

$$0<\frac{1}{\sigma}-1\leq\frac{|u_{n+1}^{-}-u_{n+2}^{-}|}{|u_{n+2}^{-}-c|}\leq\frac{K}{\frac{b}{K(1-b)}-1},$$

which becomes arbitrarily close to 0 if b is close to 1. Hence, $b < 1$ uniformly. The last statement of the lemma can be proved similarly. □

We are now ready to prove the main propositions of this section.

Proposition 6.2.7 *Let $f \in \mathcal{I}^3$. The mesh of the dynamical partitions satisfy*

$$\delta(n)\longrightarrow 0$$

when $n \to \infty$ and

$$\delta(\alpha_n)\longrightarrow 0$$

exponentially fast when $n \to \infty$.

Proof From Lemma 4.2.6 we know that $\delta(n) \to 0$. For the second statement without loss of generality we may assume $R^{\alpha_n} f \in \mathcal{D}^3_-$. We claim that there exists $b < 1$ such that for all $n \in \mathbb{N}$ and $k < r_{\alpha_n}$,

$$\frac{|U_{n,a_{n+1}+1,k}|}{|f^k(R_{\alpha_n})|} \le b.$$

Moreover if $a_{n+1} \ge 2$ or if $a_{n+1} = a_{n+2} = 1$ then

$$\frac{|U_{n,i,k}|}{|f^k(R_{\alpha_n})|} \le b, \quad \forall\, 1 \le i \le a_{n+1} + 1. \tag{6.2.4}$$

To prove this claim notice that by the first statement of Lemma 6.2.6 we have a constant $\sigma < 1$ such that for all $n \in \mathbb{N}$,

$$\frac{|U_{n,a_{n+1}+1}|}{|R_{\alpha_n}|} \le \sigma.$$

Hence

$$|R_{\alpha_n} \backslash U_{n,a_{n+1}+1}| \ge (1-\sigma)|R_{\alpha_n}|. \tag{6.2.5}$$

Assume by contradiction that for any $b < 1$ there exist n and k such that

$$\frac{|U_{n,a_{n+1}+1,k}|}{|f^k(R_{\alpha_n})|} > b,$$

then

$$|f^k(R_{\alpha_n} \backslash U_{n,a_{n+1}+1})| = |f^k(R_{\alpha_n}) \backslash U_{n,a_{n+1}+1,k}| < (1-b)|f^k(R_{\alpha_n})| \tag{6.2.6}$$

is a very small portion of $f^k(R_{\alpha_n})$. By the mean value theorem there exists $\theta_0, \theta \in R_{\alpha_n}$ such that

$$Df^k(\theta_0) = \frac{|f^k(R_{\alpha_n})|}{|R_{\alpha_n}|},$$

and

$$\begin{aligned} Df^k(\theta) &= \frac{|f^k(R_{\alpha_n} \backslash U_{n,a_{n+1}+1})|}{|R_{\alpha_n} \backslash U_{n,a_{n+1}+1}|} \\ &= \frac{|f^k(R_{\alpha_n} \backslash U_{n,a_{n+1}+1})|}{|f^k(R_{\alpha_n})|} \cdot \frac{|f^k(R_{\alpha_n})|}{|R_{\alpha_n}|} \cdot \frac{|R_{\alpha_n}|}{|R_{\alpha_n} \backslash U_{n,a_{n+1}+1}|} \\ &\le (1-b) \cdot Df^k(\theta_0) \cdot \frac{1}{1-\sigma}. \end{aligned}$$

Then

$$\mathrm{Dist}(f^k|_{R_{\alpha_n}}) \geq \left|\log \frac{Df^k(\theta)}{Df^k(\theta_0)}\right| \geq |\log(1-b) - \log(1-\sigma)|.$$

But by Lemma 4.6.1 $\mathrm{Dist}(f^k|_{R_{\alpha_n}}) \leq K$ for some constant K. Choose $b < 1$ such that $|\log(1-b)| > K + |\log(1-\sigma)|$ and we get a contradiction. This proves the first statement of the claim.

For the second statement of the claim, observe that by Lemma 6.2.6 there exists $\sigma < 1$ such that for all $n \in \mathbb{N}$ with $a_{n+1} \geq 2$ or $a_{n+1} = a_{n+2} = 1$,

$$\frac{|U_{n,i}|}{|R_{\alpha_n}|} \leq \sigma, \quad \forall\, 1 \leq i \leq a_{n+1} + 1.$$

Hence at least one component of $R_{\alpha_n} \backslash U_{n,i}$ cannot be too small. Using the similar argument as before, assume $|U_{n,i,k}|$ is almost as large as $|f^k(R_{\alpha_n})|$ for some n, k then we get a contradiction with the distortion bound. This finishes the proof of the claim.

Note that for $n \in \mathbb{N}$ with $\mathit{fast}R^n f \in \mathcal{D}^3_+$ an analogous claim holds. Using the claim and Remark 6.2.1 we have

- If $a_{n+1} \neq 1$ then

$$\delta_+(\alpha_{n+1}) \leq b \cdot \delta_+(\alpha_n)$$
$$\delta_-(\alpha_{n+1}) \leq \max\{\, b \cdot \delta_+(\alpha_n),\ \delta_-(\alpha_n)\,\}$$

- If $a_{n+1} = 1$ and $a_{n+2} \neq 1$ then

$$\delta_+(\alpha_{n+1}) \leq b \cdot \delta_+(\alpha_n)$$
$$\delta_-(\alpha_{n+1}) \leq \max\{\, \delta_+(\alpha_n),\ \delta_-(\alpha_n)\,\}$$

- If $a_{n+1} = 1$ and $a_{n+2} = 1$ then

$$\delta_+(\alpha_{n+1}) \leq b \cdot \delta_+(\alpha_n)$$
$$\delta_-(\alpha_{n+1}) \leq \max\{\, b \cdot \delta_+(\alpha_n),\ \delta_-(\alpha_n)\,\}$$

- If $a_{n+2} \neq 1$ then

$$\delta_+(\alpha_{n+2}) \leq \max\{\, \delta_+(\alpha_{n+1}),\ b \cdot \delta_-(\alpha_{n+1})\,\}$$
$$\delta_-(\alpha_{n+2}) \leq b \cdot \delta_-(\alpha_{n+1})$$

- If $a_{n+2} = 1$ then

$$\delta_+(\alpha_{n+2}) \leq \max\{\,\delta_+(\alpha_{n+1}),\ \delta_-(\alpha_{n+1})\,\}$$
$$\delta_-(\alpha_{n+2}) \leq b \cdot \delta_-(\alpha_{n+1})$$

This implies, by iterating two steps, and using the claim again,

- If $a_{n+2} \neq 1$ then

$$\delta_+(\alpha_{n+2}) \leq b \cdot \max\{\,\delta_+(\alpha_n),\ \delta_-(\alpha_n)\,\}$$
$$\delta_-(\alpha_{n+2}) \leq b \cdot \max\{\,\delta_+(\alpha_n),\ \delta_-(\alpha_n)\,\}$$

- If $a_{n+2} = 1$ and $a_{n+1} \neq 1$ then

$$\delta_+(\alpha_{n+2}) \leq \max\{\,b \cdot \delta_+(\alpha_n),\ \delta_-(\alpha_n)\,\}$$
$$\delta_-(\alpha_{n+2}) \leq b \cdot \max\{\,\delta_+(\alpha_n),\ \delta_-(\alpha_n)\,\}$$

- If $a_{n+2} = 1$ and $a_{n+1} = 1$ then

$$\delta_+(\alpha_{n+2}) \leq \max\{\,b \cdot \delta_+(\alpha_n),\ \delta_-(\alpha_n)\,\}$$
$$\delta_-(\alpha_{n+2}) \leq b \cdot \max\{\,\delta_+(\alpha_n),\ \delta_-(\alpha_n)\,\}.$$

This implies in turn, by iterating four steps,

- If $a_{n+2} \neq 1$ or $a_{n+4} \neq 1$ then

$$\delta(\alpha_{n+4}) \leq b \cdot \delta(\alpha_n).$$

- If $a_{n+2} = 1$ and $a_{n+4} = 1$ then

$$\delta_+(\alpha_{n+4}) \leq \max\{\,b \cdot \delta_+(\alpha_{n+2}),\ \delta_-(\alpha_{n+2})\,\}$$
$$\delta_-(\alpha_{n+4}) \leq b \cdot \max\{\,\delta_+(\alpha_{n+2}),\ \delta_-(\alpha_{n+2})\,\},$$

  which implies

$$\delta_+(\alpha_{n+4}) \leq \max\{\,b \cdot \delta_+(\alpha_{n+2}),\ b \cdot \delta(\alpha_n)\,\}$$
$$\delta_-(\alpha_{n+4}) \leq b \cdot \delta(\alpha_n).$$

Both situations together imply that for every $n \in \mathbb{N}$

$$\delta(\alpha_{n+4}) \leq b \cdot \delta(\alpha_n).$$

This means that $\delta(\alpha_n)$ decays exponentially. □

Proposition 6.2.8 *Let $f \in \mathcal{I}^3$. Then*

$$|z(n)| \longrightarrow 0$$

when $n \to \infty$ and

$$|z(\alpha_n)| \longrightarrow 0$$

exponentially fast when $n \to \infty$.

Proof For the second statement observe that by Lemmas 6.2.2 and 6.2.6 there exists a constant $b < 1$ such that for all $n \in \mathbb{N}$

$$|z(\alpha_{n+1})| \leq b|z(\alpha_n)| + C\delta(\alpha_n).$$

From this estimates and from Proposition 6.2.7 $|z(\alpha_n)| \to 0$ exponentially fast. The first statement follows from Lemma 6.2.2, Proposition 6.2.7 and the fact that $|z(\alpha_n)| \to 0$ exponentially fast. □

Chapter 7
Exponential Convergence

In the previous chapters we prepared the stage to prove the essential and most delicate ingredient of renormalization theory, namely consecutive renormalizations of circle diffeomorphisms converge exponentially fast to the set of rotations. Unfortunately the picture is more complicated than what one would wish for; in particular there are examples where one does not see convergence at all. These examples will be discussed in Chap. 9.

The examples in Chap. 9 are maps which have the characteristic property that the sequence a_n converges very fast to infinity. To assure convergence we restrict ourself to maps with bounded combinatorics, i.e. the numbers a_n are bounded. Section 7.2 is dedicated to prove the exponential convergence of all renormalizations of circle diffeomorphisms with bounded combinatorics to the set of rotations.

The restriction to maps with bounded combinatorics is not needed if one only considers fast renormalizations. Indeed in Sect. 7.1 we establish that the subsequence defined by consecutive fast renormalizations converges exponentially fast to the rotations with no restriction on the rotation number.

7.1 Fast Renormalization

Let $\mathcal{W} = \{r_\rho | \rho \in [0, 1)\} \subset \mathcal{D}^3$ be the set of rotations. In this section we will show that

$$fastR^n f \to \mathcal{W}$$

in the d_1 topology with exponential rate.

Y. Dong et al., *One-Dimensional Dynamics*, Lecture Notes of the Unione Matematica Italiana 30, https://doi.org/10.1007/978-3-031-94882-4_7

Theorem 7.1.1 *Let $f \in \mathcal{I}^3$ then there exist constants $C > 0$ and $0 < \rho < 1$ such that*

$$d_1(fastR^n f, \mathcal{W}) \leq C\rho^n. \tag{7.1.1}$$

Notice that, by definition

$$\begin{aligned} d_1(fastR^n f, \mathcal{W}) \\ &= \min_{t\in[0,1)} (\max\{dist_{\mathcal{C}^1}(N(R^{\alpha_n} f)_\pm, \mathrm{id}), \left|c_{R^{\alpha_n} f} - (1-t)\right|, \left|v_{R^{\alpha_n} f} - t\right|\}) \\ &\leq \max\{dist_{\mathcal{C}^1}(N(R^{\alpha_n} f)_\pm, \mathrm{id}), \left|c_{R^{\alpha_n} f} - (1 - v_{R^{\alpha_n} f})\right|\}. \end{aligned} \tag{7.1.2}$$

Compare this with the proof of Proposition 5.3.1. Using the results in the previous two chapters, we first prove that the renormalizations of a circle diffeomorphism f become linear, i.e. $R^n f$ converge to affine maps. This implies that the first term in (7.1.2) converges to 0.

Proposition 7.1.2 *Let $f \in \mathcal{I}^3$ then*

$$d_1(R^n f, \mathcal{A}) \longrightarrow 0$$

when $n \to \infty$ and

$$d_1(fastR^n f, \mathcal{A}) \longrightarrow 0$$

exponentially fast when $n \to \infty$.

Proof Recall that $K(n) = C(|z(n)|+\delta(n))$, where C is a sufficiently large constant. Notice that by Proposition 5.1.12, Lemma 5.2.2 and (6.1.2), we have

$$d_1(R^n f, \mathcal{A}) \leq d_1(R^n f, m_n) + d_1(m_n, \mathcal{A}) \leq C\delta(n) + C|w(n)| \leq K(n),$$

and in particular,

$$d_1(fastR^n f, \mathcal{A}) \leq K(\alpha_n) = C(|z(\alpha_n)| + \delta(\alpha_n)).$$

The proposition follows by applying Propositions 6.2.7 and 6.2.8.

□

We are now ready to prove Theorem 7.1.1.

Proof (Proof of Theorem 7.1.1)
Proposition 7.1.2 yields that the first term in (7.1.2) decays exponentially,

$$dist_{\mathcal{C}^1}(N(R^{\alpha_n} f)_\pm, \mathrm{id}) \leq d_1(fastR^n f, \mathcal{A}) \leq C\rho^n.$$

We still need to show that the second term of (7.1.2) goes to 0 exponentially fast. Because $|s(\alpha_n)|$ is bounded (see Proposition 6.2.5), we have

$$\left|c_{R^{\alpha_n} f} - (1 - v_{R^{\alpha_n} f})\right| = c_{R^{\alpha_n} f}|1 - d_-(\alpha_n)| \le C|s_-(\alpha_n)| \le C|s(\alpha_n)|.$$

Hence at this point to prove Theorem 7.1.1 it suffices to show for some constant $C_f > 0$ and $\rho_0 < 1$,

$$|s(\alpha_n)| \le C_f \rho_0^n. \tag{7.1.3}$$

By Propositions 6.2.7 and 6.2.8, there exists $\rho_1 \in (0, 1)$ and $K_1 > 0$ such that $K(\alpha_n) \le K_1 \rho_1^n$ for all $n \in \mathbb{N}$. By Lemma 6.2.2, for all n and $0 \le i \le a_{n+1}$ we have

$$\begin{aligned}
K(\alpha_n + i) &= C(|z(\alpha_n + i)| + \delta(\alpha_n + i)) \\
&\le C((\sigma_{n,i}|z(\alpha_n)| + C\delta(\alpha_n)) + \delta(\alpha_n)) \\
&\le C'C(|z(\alpha_n)| + \delta(\alpha_n)) = C'K(\alpha_n) \le C'K_1\rho_1^n.
\end{aligned}$$

Let $K = C'K_1$ and $\rho = \max\{\frac{11}{12}, \rho_1\}$, then we have for all $n \in \mathbb{N}$,

$$K(\alpha_n + i) \le K\rho^n, \quad \forall\, 0 \le i \le a_{n+1}. \tag{7.1.4}$$

Let $L > 0$ be a sufficiently large constant such that $K \le \frac{1}{48}L$. Assume by induction that $|s(\alpha_n)| \le L\rho^n$ and we will show that $|s(\alpha_{n+1})| \le L\rho^{n+1}$. We consider the case *fast*$R^n f \in \mathcal{D}^3_-$, the other case can be treated similarly.

Let Δ denote the two cones in the coordinate plane of (s_+, s_-) formed by the lines

$$s_- = -3s_+ \quad \text{and} \quad s_- = -s_+,$$

namely

$$s = (s_+, s_-) \in \Delta \quad \Leftrightarrow \quad |s_+| < |s_-| < 3|s_+|,\ s_+s_- < 0. \tag{7.1.5}$$

When we apply the renormalization operator to any (s_+, s_-) in the area Δ, the L^1 norm $|s|$ is contracted by ratio $\frac{3}{4}$ up to an error term. In fact, for any $0 \le i \le a_{n+1} - 1$ such that $s(\alpha_n + i) = (s_+(\alpha_n + i), s_-(\alpha_n + i)) \in \Delta$, note that $R^{\alpha_n + i} \in \mathcal{D}^3_-$, by Proposition 6.1.3 and (7.1.4) we have

$$\begin{cases}
s_+(\alpha_n + i) - K\rho^n \le s_+(\alpha_n + i + 1) \le s_+(\alpha_n + i) + K\rho^n, \\
s_+(\alpha_n + i) + s_-(\alpha_n + i) - K\rho^n \le s_-(\alpha_n + i + 1) \le s_+(\alpha_n + i) \\
\qquad + s_-(\alpha_n + i) + K\rho^n.
\end{cases}$$

Hence,

$$
\begin{aligned}
|s(\alpha_n+i+1)| &= |s_+(\alpha_n+i+1)-s_-(\alpha_n+i+1)| \le |s_-(\alpha_n+i)|+2K\rho^n\\
&= \frac{3}{4}|s_-(\alpha_n+i)|+\frac{1}{4}|s_-(\alpha_n+i)|+2K\rho^n\\
&\le \frac{3}{4}|s_-(\alpha_n+i)|+\frac{1}{4}\cdot 3|s_+(\alpha_n+i)|+2K\rho^n\\
&= \frac{3}{4}|s(\alpha_n+i)|+2K\rho^n,
\end{aligned}
\tag{7.1.6}
$$

where the last inequality is obtained from the condition in (7.1.5).

Suppose $s(\alpha_n+i)\in\Delta$ for all $0\le i\le a_{n+1}-1$, then from the estimate (7.1.6) we have

$$
\begin{aligned}
|s(\alpha_{n+1})| &\le \frac{3}{4}|s(\alpha_n+a_{n+1}-1)|+2K\rho^n\\
&\le \frac{3}{4}\left(\frac{3}{4}|s(\alpha_n+a_{n+1}-2)|+2K\rho^n\right)+2K\rho^n\\
&\le \cdots\\
&\le \left(\frac{3}{4}\right)^{a_{n+1}}\cdot|s(\alpha_n)|+\left(1+\frac{3}{4}+\cdots+\left(\frac{3}{4}\right)^{a_{n+1}-1}\right)\cdot 2K\rho^n\\
&\le \frac{3}{4}|s(\alpha_n)|+8K\rho^n.
\end{aligned}
\tag{7.1.7}
$$

Using the induction hypothesis and by our choice of the constants L and ρ we get

$$
|s(\alpha_{n+1})| \le \frac{3}{4}L\rho^n+8\cdot\frac{1}{48}L\rho^n=\frac{11}{12}L\rho^n\le L\rho^{n+1}.
$$

Hence when all the renormalizations are in the area Δ we have the exponential decay of $|s(\alpha_n)|$.

Suppose that there exists $0\le i\le a_{n+1}-1$ such that $s(\alpha_n+i)\notin\Delta$. Let i_0 denote the smallest such i. In particular, $s(\alpha_n+i)\in\Delta$ for $i=0,1,\ldots,i_0-1$. Then using (7.1.6) for these i_0 steps we compute similarly as in (7.1.7) to get

$$
|s(\alpha_n+i_0)| \le \left(\frac{3}{4}\right)^{i_0}|s(\alpha_n)|+8K\rho^n\le L\rho^n+8K\rho^n.
$$

Observe that $R^{\alpha_n+i_0}f\in\mathcal{D}^3_-$ and $s(\alpha_n+i_0)\notin\Delta$ imply

$$
|s_+(\alpha_n+i_0)|\le\frac{1}{3}|s_-(\alpha_n+i_0)|.
$$

By Lemma 6.2.3 we have

$$|s_+(\alpha_{n+1})| \leq |s_+(\alpha_n + i_0)| + K\rho^n.$$

Since $R^{\alpha_{n+1}} f \in \mathcal{D}^3_+$, by Lemma 6.1.2 we have

$$|s(\alpha_{n+1})| = |s_+(\alpha_{n+1})| + |s_-(\alpha_{n+1})| \leq 2|s_+(\alpha_{n+1})|.$$

Combining these last four inequalities, we have

$$\begin{aligned}
|s(\alpha_{n+1})| &\leq 2\left(|s_+(\alpha_n + i_0)| + K\rho^n\right) \\
&\leq 2\left(\frac{1}{3}|s_-(\alpha_n + i_0)| + K\rho^n\right) \\
&\leq 2\left(\frac{1}{3}(L\rho^n + 8K\rho^n) + K\rho^n\right) \\
&\leq \frac{2}{3}L\rho^n + \frac{22}{3}\cdot\frac{1}{48}L\rho^n < \frac{11}{12}L\rho^n \leq L\rho^{n+1}.
\end{aligned}$$

This concludes the proof of (7.1.3) and thus the proof of Theorem 7.1.1.

□

The rate of convergence ρ from Theorem 7.1.1 comes from Proposition 6.2.7 and Proposition 6.2.8. Note that the contracting constant $b < 1$ in the proof of Proposition 6.2.7 is determined by the constant σ in Lemma 6.2.6, which comes from the a priori bound $\frac{1}{K_f} \leq DfastR^n f \leq K_f$. Here the constant K_f depends on f, and consequently the rate of convergence ρ depends of f. However we have already proved the exponential convergence of $R^{\alpha_n} f$ to rotation in the d_1 topology, which implies the derivative of $fastR^n f$ becomes close to 1 when n is large. Using this we can prove the existence of a universal rate of convergence ρ_0, namely as a corollary of Theorem 7.1.1 we have the following theorem.

Theorem 7.1.3 *There exists $\rho_0 < 1$ such that the following holds. Let $f \in \mathcal{I}^3$ then there exists a constant $C_f > 0$ such that*

$$d_1(fastR^n f, \mathcal{W}) \leq C_f \rho_0^n.$$

Proof Let $f \in \mathcal{I}^3$. Note that for any $x \in [c_n, 1]$ and its corresponding $\theta \in R_{\alpha_n}$,

$$\begin{aligned}
|\log D(R^{\alpha_n})f(x)| &= \left|\log DN(R^{\alpha_n} f)_+(\theta) + \log \frac{v_{\alpha_n}}{1 - c_{\alpha_n}}\right| \\
&\leq |\log DN(R^{\alpha_n} f)_+(\theta)| + |s_+(\alpha_n)| \\
&\leq C dist_{C^1}(N(R^{\alpha_n} f)_+, \mathrm{id}) + |s_+(\alpha_n)|.
\end{aligned}$$

From the proof of Theorem 7.1.1, both $dist_{\mathcal{C}^1}(N(R^{\alpha_n} f)_+, \mathrm{id})$ and $|s_+(\alpha_n)|$ goes to 0 exponentially fast. There exists a sufficiently large integer $N_+ \in \mathbb{N}$ such that for all $n \geq N_+$

$$dist_{\mathcal{C}^1}(N(R^{\alpha_n} f)_+, \mathrm{id}), |s_+(\alpha_n)| \leq \frac{\log 2}{C+10},$$

then

$$\frac{1}{2} \leq \exp\left(-(C+1)\frac{\log 2}{C+10}\right) \leq |DR^{\alpha_n} f(x)| \leq \exp\left((C+1)\frac{\log 2}{C+10}\right) \leq 2$$

for all $n \geq N_+$. Apply the similar argument to $x \in [0, c_n]$ and we get an integer $N_- \in \mathbb{N}$ such that

$$\frac{1}{2} \leq |DR^{\alpha_n} f(x)| \leq 2, \quad \forall n \geq N_-, \forall x \in [0, c_n].$$

Let $N_f := \max\{N_+, N_-\}$ and we get a uniform bound on the derivative of the fast renormalizations of the given map f which holds for all but the first N_f fast renormalizations. As we discussed above, using this bound we will get a constant $\rho_0 < 1$ such that (7.1.1) holds for all but finitely many steps of fast renormalizations of f. Hence by choosing a sufficiently large constant C_f we can ensure that (7.1.1) with $\rho = \rho_0$ holds for all fast renormalizations of f. □

7.2 Renormalization and Bounded Combinatorics

As mentioned at the beginning of this chapter, in general the renormalizations of a map $f \in \mathcal{I}^3$ do not necessarily converge to the set of rotations. In this section we consider the generalized $\mathcal{C}^3$ circle diffeomorphisms with bounded combinatorics. For a given integer $A > 0$, we define the following set

$$\mathcal{I}^3_A := \{ f \in \mathcal{I}^3 \mid a_n(f) \leq A, \ \forall n \geq 1 \} \subset \mathcal{I}^3. \tag{7.2.1}$$

The renormalizations of such maps indeed converge to the set of rotations.

Theorem 7.2.1 *Let $f \in \mathcal{I}^3_A$ for some $A > 0$. Then there exist constants $C > 0$ and $\xi \in (0, 1)$ such that*

$$d_1(R^n f, \mathcal{W}) \leq C\xi^n.$$

The proof of Theorem 7.2.1 relies on an essential feature of maps with bounded rotation numbers, namely *bounded geometry*, i.e. the scaling ratio σ is bounded from above and from below. This does not hold for general combinatorics. Bounded geometry allows to prove the exponential decay of the error terms without the need of introducing the notion of fast renormalization.

We follow the same structure as the proof of the exponential convergence of fast renormalizations. Compare the following proposition with Proposition 6.2.5.

Proposition 7.2.2 *Let $f \in \mathcal{I}^3_A$. Then there exists a constant $K_f > 1$ such that*

$$\frac{1}{K_f} \le DR^n f \le K_f.$$

Proof Similar as in the proof of Proposition 6.2.5, as a consequence of the bounded distortion (Lemma 4.6.1) it suffices to show that $|s(n)|$ is bounded for all n.

Recall that $K(n) = C_0(|z(n)| + \delta(n))$ where $C_0 > 0$ is a constant such that Proposition 6.1.3 holds, and $K = \sup_n K(n) < \infty$, see (6.1.3). By Lemma 6.2.4, for a given $n \in \mathbb{N}$, if all steps $i = 0, 1, \ldots, n-1$ satisfy

$$\begin{aligned} |s_+(i)| &\ge 2K(i) \quad \text{if } R^i f \in \mathcal{D}^3_-, \\ |s_-(i)| &\ge 2K(i) \quad \text{if } R^i f \in \mathcal{D}^3_+, \end{aligned} \tag{7.2.2}$$

then

$$|s(n)| \le |s(n-1)| \le \cdots \le |s(0)|$$

is bounded.

Suppose now that (7.2.2) does not hold for some $m \le n-1$. Let m denote the largest such number, so that (7.2.2) holds for all steps $m+1, \cdots, n-1$ and we have

$$|s(n)| \le |s(n-1)| \le \cdots \le |s(m)|. \tag{7.2.3}$$

We consider the case $R^m f \in \mathcal{D}^3_-$. In particular, $|s_+(m)| < 2K(m) \le 2K$. Note that after at most A steps of renormalization the map goes to $\mathcal{D}^3_+$. Let $j > 0$ be the smallest number such that $R^{m+j} f \in \mathcal{D}^3_+$, then $j \le A$, and by Lemma 6.1.2 we have

$$|s_-(m+j)| < |s_+(m+j)|. \tag{7.2.4}$$

Observe that by our choice of j, for all $i = 0, 1, \ldots, j-1$ we have $R^{m+i} f \in \mathcal{D}^3_-$. By Proposition 6.1.3, for $0 \le i \le j-1$

$$\begin{aligned} |s_+(m+i+1)| &\le |s_+(m+i)| + K(m+i) \le |s_+(m+i)| + K \\ &\le \cdots \le |s_+(m)| + iK \le |s_+(m)| + AK. \end{aligned} \tag{7.2.5}$$

In particular,

$$|s_+(m+j)| \le |s_+(m)| + AK. \tag{7.2.6}$$

Using Proposition 6.1.3 again and (7.2.5) we have

$$\begin{aligned}|s_-(m+j)| &\geq |s_+(m+j-1)+s_-(m+j-1)-K|\\ &\geq |s_-(m+j-1)| - (|s_+(m)|+AK) - K\\ &\geq \cdots \geq |s_-(m)| - A(|s_+(m)|+AK) - AK.\end{aligned}$$

This together with (7.2.6), (7.2.4) and the fact that $|s_+(m)| < 2K$ give us

$$|s_-(m)| \leq A(|s_+(m)|+AK) + AK + |s_+(m)| + AK \leq (A^2+4A+2)K.$$

By (7.2.3) and Lemma 6.1.2,

$$|s(n)| \leq |s(m)| = |s_+(m)| + |s_-(m)| \leq 2|s_-(m)| \leq 2(A^2+4A+2)K.$$

We have proved that

$$|s(n)| \leq \max\{\, 2(A^2+4A+2)K,\ |s(0)|\,\}$$

is bounded. □

Recall the definition of scaling ratio,

$$\sigma(n) = \begin{cases} \frac{|R_{n+1}|}{|R_n|} & \text{if } R^n f \in \mathcal{D}^3_- \\ \frac{|L_{n+1}|}{|L_n|} & \text{if } R^n f \in \mathcal{D}^3_+ \end{cases}.$$

Using the bound for derivatives, we now prove that, for any map $f \in \mathcal{I}^3_A$ the scaling ratio $\sigma(n)$ cannot be arbitrarily close to 0 or 1, which lead to prove that the mesh of the dynamical partition, $\delta(n)$, contracts by a constant strictly less than 1 at each step. This is indeed the main difference between the proof for maps with bounded combinatorics and for general $\mathcal{I}^3$ maps. In particular, compare the following Proposition with Lemma 6.2.6.

Proposition 7.2.3 *Let $f \in \mathcal{I}^3_A$. Then there exist constants $\sigma_1, \sigma_2 \in (0,1)$ such that*

$$0 < \sigma_1 \leq \sigma(n) \leq \sigma_2 < 1, \quad \forall\, n \in \mathbb{N}.$$

Proof We first prove the existence of the upper bound. Suppose $R^n f \in \mathcal{D}^3_-$. Let J_0 denote the interval $R_n \backslash R_{n+1}$ and we show that J_0 is not a too small portion of R_n. Similarly as before, let $j > 0$ denote the smallest number such that $R^{n+j} f \in \mathcal{D}^3_+$, and we have $j \leq A$. For $i = 0, 1, \ldots, j$, let $J_i := (f^{r_n})^i(J_0)$. By Proposition 7.2.2, there exists a constant $K_f > 1$ such that

$$\frac{1}{K_f} \leq D(f^{r_n}|_{R_n}) \leq K_f.$$

Hence, by the mean value theorem we have

$$\frac{|J_i|}{|J_{i-1}|} \leq K_f, \quad \forall i = 1, \ldots, j.$$

As a consequence,

$$|J_i| \leq K_f^i |J_0| \leq K_f^A |J_0|, \quad \forall i = 0, 1, \ldots, j.$$

Observe that by our choice of j, we have

$$|R_n| \leq |J_0| + |J_1| + \cdots |J_j| \leq j \cdot K_f^A |J_0| \leq A K_f^A |J_0|.$$

Hence,

$$\frac{|J_0|}{|R_n|} \geq \frac{1}{A K_f^A}. \tag{7.2.7}$$

Let $\sigma_2 := 1 - \frac{1}{AK_f^A} < 1$ then we have

$$\sigma(n) = \frac{|R_{n+1}|}{|R_n|} = 1 - \frac{|J_0|}{|R_n|} \leq \sigma_2.$$

The case $R^n f \in \mathcal{D}_+^3$ can be treated similarly, hence there exists a constant $\sigma_2 \in (0, 1)$ such that

$$\sigma(n) \leq \sigma_2 < 1, \quad \forall n \in \mathbb{N}. \tag{7.2.8}$$

We use (7.2.8) to prove the existence of the lower bound. We consider the case $R^n f \in \mathcal{D}_-^3$. Let $j > 0$ be the smallest positive integer such that $R^{n+j} f \in \mathcal{D}_+^3$ as before. Observe that

$$\left| f^{r_{n+j}}(R_{n+j}) \right| = |L_{n+j} \backslash L_{n+j+1}|. \tag{7.2.9}$$

By Proposition 7.2.2 and the mean value theorem,

$$\left| f^{r_{n+j}}(R_{n+j}) \right| \leq K_f |R_{n+j}| \leq K_f |R_{n+1}|. \tag{7.2.10}$$

Using the upper bound (7.2.8),

$$\frac{|L_{n+j+1}|}{|L_{n+j}|} = \sigma(n + j) \leq \sigma_2.$$

Hence

$$|L_{n+j} \backslash L_{n+j+1}| = |L_{n+j}| - |L_{n+j+1}| \geq (1 - \sigma_2)|L_{n+j}| = (1 - \sigma_2)|L_n|. \tag{7.2.11}$$

Moreover, notice that $|J_0| = |f^{l_n}(L_n)|$, using (7.2.7) and Proposition 7.2.2 we have

$$\frac{1}{AK_f^A} \leq \frac{|f^{l_n}(L_n)|}{|R_n|} \leq \frac{K_f|L_n|}{|R_n|}. \tag{7.2.12}$$

Combining (7.2.10), (7.2.9), (7.2.11), and (7.2.12) we get

$$K_f|R_{n+1}| \geq (1 - \sigma_2)\frac{1}{AK_f^{A+1}}|R_n|.$$

Hence,

$$\sigma(n) = \frac{|R_{n+1}|}{|R_n|} \geq \sigma_1 := \frac{1 - \sigma_2}{AK_f^{A+2}} > 0.$$

The lower bound of $\sigma(n)$ for the case $R^n f \in \mathcal{D}_+^3$ can be obtained similarly. □

We can now prove the exponential decay of $\delta(n)$ and $|z(n)|$, namely the analogies of Proposition 6.2.7 and Proposition 6.2.8.

Proposition 7.2.4 *Let $f \in \mathcal{I}_A^3$. Then there exist constants $C > 0$ and $\xi \in (0, 1)$ such that*

$$\delta(n) \leq C\xi^n, \quad \forall n \in \mathbb{N}.$$

Proof Suppose $R^n f \in \mathcal{D}_-^3$. Then by Proposition 7.2.3,

$$\frac{|R_{n+1}|}{|R_n|} \leq \sigma_2 < 1.$$

Similar as in the proof of Proposition 6.2.7, as a consequence of the bounded distortion, there exists a constant $b_2 \in (0, 1)$ such that for all $0 \leq k \leq r_n - 1$,

$$\frac{|f^k(R_{n+1})|}{|f^k(R_n)|} \leq b_2 < 1,$$

where r_n is the right first return time, see Lemma 4.1.6. This implies that

$$\delta_+(n + 1) \leq b_2 \cdot \delta_+(n), \quad \text{if } R^n f \in \mathcal{D}_-^3. \tag{7.2.13}$$

On the other hand, by Proposition 7.2.3 we also have

$$\frac{|J_0|}{|R_n|} \leq 1 - \sigma_1 < 1,$$

where J_0 denote the interval $R_n \backslash R_{n+1}$. Using similar arguments as before we can prove that there exists a constant $b_1 \in (0, 1)$ such that for all $0 \le k \le r_n - 1$,

$$\frac{|f^{l_n+k}(L_n)|}{|f^k(R_n)|} = \frac{|f^k(J_0)|}{|f^k(R_n)|} \le b_1 < 1.$$

Hence if $R^n f \in \mathcal{D}_-^3$ we have

$$\delta_-(n+1) \le \max\{\delta_-(n),\ b_1 \cdot \delta_+(n)\}.$$

Similarly we can show that if $R^n f \in \mathcal{D}_+^3$ then

$$\delta_-(n+1) \le b_2 \cdot \delta_-(n), \tag{7.2.14}$$

and

$$\delta_+(n+1) \le \max\{\delta_+(n),\ b_1 \cdot \delta_-(n)\}. \tag{7.2.15}$$

Since for all $n \in \mathbb{N}$ the dynamical partition $\mathscr{P}_{n+1}$ is a refinement of $\mathscr{P}_n$, we have

$$\delta(n+1) \le \delta(n), \quad \forall n \in \mathbb{N}. \tag{7.2.16}$$

Let $b := \max\{b_1, b_2\} < 1$. For any given $m \in \mathbb{N}$, we consider the case $R^m f \in \mathcal{D}_-^3$. As before let $j \in \mathbb{N}^+$ be the smallest number such that $R^{m+j} f \in \mathcal{D}_+^3$. Since $R^m f \in \mathcal{D}_-^3$, by (7.2.13) and (7.2.16) we have

$$\delta_+(m+j) \le b \cdot \delta_+(m)$$

and

$$\delta_-(m+j) \le \delta(m+j) \le \delta(m).$$

Since $R^{m+j} f \in \mathcal{D}_+^3$, by (7.2.15) and (7.2.14) we have

$$\begin{aligned} \delta_+(m+j+1) &\le \max\{\delta_+(m+j),\ b \cdot \delta_-(m+j)\} \\ &\le \max\{b \cdot \delta_+(m),\ b \cdot \delta(m)\} \\ &\le b \cdot \delta(m), \end{aligned}$$

and

$$\delta_-(m+j+1) \le b \cdot \delta_-(m+j) \le b \cdot \delta(m).$$

Hence (recall that $j \le A$),

$$\delta(m+A+1) \le \delta(m+j+1) \le b \cdot \delta(m).$$

The case $R^m f \in \mathcal{D}_+^3$ can be treated similarly, and we get

$$\delta(m+A+1) \le b \cdot \delta(m), \quad \forall m \in \mathbb{N}.$$

Suppose $n = k(A+1) + l$ for $k, l \in \mathbb{N}$ with $l < A+1$. Then $k \le \frac{n}{A+1}$ and we have

$$\delta(n) \le b^k \cdot \delta(l) \le b^k \le b^{\frac{n}{A+1}}.$$

Set $\xi := b^{\frac{1}{A+1}} < 1$ and we have $\delta(n) \le \xi^n$. □

Proposition 7.2.5 *Let $f \in \mathcal{I}_A^3$. Then there exist constants $C > 0$ and $\xi \in (0,1)$ such that*

$$|z(n)| \le C\xi^n, \quad \forall n \in \mathbb{N}.$$

Proof By Propositions 6.1.3, 7.2.3, and 7.2.4, there exist constants $C > 0$, $\sigma < 1$ and $\zeta < 1$ such that for all $n \in \mathbb{N}$,

$$|z(n+1)| \le \sigma|z(n)| + C\zeta^n.$$

Hence

$$\begin{aligned}
|z(n)| &\le \sigma|z(n-1)| + C\zeta^{n-1} \\
&\le \sigma\left(\sigma|z(n-2)| + C\zeta^{n-2}\right) + C\zeta^{n-1} \\
&\le \cdots \\
&\le \sigma^n|z(0)| + C(\zeta^{n-1} + \sigma\zeta^{n-2} + \cdots + \sigma^{n-1})
\end{aligned}$$

Choose $\xi \in (\max\{\sigma, \zeta\}, 1)$ and we have

$$|z(n)| \le C\xi^n.$$

□

We are now ready to prove Theorem 7.2.1.

Proof of Theorem 7.2.1 Similar as in the proof of Theorem 7.1.1, by definition of the d_1 distance we are left to show that $|c_n - (1 - v_n)|$ decay exponentially fast (see (7.1.2)). Since by Proposition 7.2.2 the slopes $|s(n)|$ are bounded for $f \in \mathcal{I}_A^3$, it suffices to show that there exist constants $C > 0$ and $\xi \in (0,1)$ such that

$$|s(n)| \le C\xi^n.$$

We proceed using induction. By Propositions 7.2.4 and 7.2.5, there exist constants $L > 0$ and $\zeta \in (0, 1)$ such that

$$K(n) = C_0(|z(n)| + \delta(n)) \leq L\zeta^n, \quad \forall n \in \mathbb{N}. \tag{7.2.17}$$

Let $C > 0$ be a constant such that $L \leq \frac{1}{2(A+2)^2}C$ and $\xi := \max\{\frac{2A+3}{2A+4}, \zeta\}$. Let C also satisfy $|s(0)| \leq C$. Assume by induction that $|s(n)| \leq C\xi^n$ and we will show that $|s(n+1)| \leq C\xi^{n+1}$.

We consider the case $R^n f \in \mathcal{D}^3_-$ and let $j \in \mathbb{N}^+$ be the smallest number such that $R^{n+j} f \in \mathcal{D}^3_+$ as before. By Proposition 6.1.3 we have

$$|s(n+1)| \leq |s_-(n)| + 2K(n) \leq |s_-(n)| + 2L\zeta^n. \tag{7.2.18}$$

As the error term is exponentially small, we want to show that $|s_-(n)|$ is a strictly smaller proportion of $|s(n)|$. We prove that since $R^n f$ needs to move to $\mathcal{D}^3_+$ in less than A steps, $|s_-(n)|$ indeed cannot be almost the same as $|s(n)|$. This is the key to the contraction of $|s(n+1)|$.

Notice that by (7.2.17), $K(k) \leq L\zeta^n$ for $k > n$. Using Proposition 6.1.3 and a computation similarly as in (7.2.5) we have for all $i = 1, \ldots, j$,

$$|s_+(n+i)| \leq |s_+(n)| + AL\zeta^n,$$

and thus

$$\begin{aligned} |s_-(n+j)| &\geq |s_-(n+j-1)| - |s_+(n+j-1)| - K(n) \\ &\geq |s_-(n+j-1)| - (|s_+(n)| + AL\zeta^n) - L\zeta^n \\ &\geq \cdots \\ &\geq |s_-(n)| - A(|s_+(n)| + AL\zeta^n) - AL\zeta^n. \end{aligned}$$

Since $R^{n+j} f \in \mathcal{D}^3_+$, by Lemma 6.1.2 we have $|s_-(n+j)| < |s_+(n+j)|$. Hence

$$\begin{aligned} |s_-(n)| &\leq A(|s_+(n)| + AL\zeta^n) + AL\zeta^n + |s_+(n)| + AL\zeta^n \\ &= (A+1)|s_+(n)| + A(A+2)L\zeta^n \\ &= (A+1)(|s(n)| - |s_-(n)|) + A(A+2)L\zeta^n. \end{aligned}$$

As a consequence,

$$|s_-(n)| \leq \frac{A+1}{A+2}C\xi^n + AL\zeta^n,$$

where we used the induction hypothesis. By (7.2.18) and our choice of C and ξ we have

$$
\begin{aligned}
|s(n+1)| &\le \frac{A+1}{A+2}C\xi^n + AL\zeta^n + 2L\zeta^n \\
&\le \left(\frac{A+1}{A+2}C + (A+2)L\right)\xi^n \\
&\le \left(\frac{A+1}{A+2}C + \frac{1}{2(A+2)}C\right)\xi^n \\
&= \frac{2A+3}{2A+4}C\xi^n \le C\xi^{n+1}.
\end{aligned}
$$

The case $R^n f \in \mathcal{D}^3_+$ can be treated similarly. This concludes the induction step. □

Remark 7.2.6 Similarly as in Theorem 7.1.3, for each constant A we can obtain a universal rate of convergence ξ_A for all $f \in \mathcal{I}^3_A$.

Chapter 8
Geometry of Circle Diffeomorphisms

In Chap. 3 we proved that any orientation preserving $\mathcal{C}^2$ diffeomorphism f having irrational rotation number ρ is conjugate to the rotation of angle ρ, i.e. there exists an homeomorphism h such that $h \circ f = r_\rho \circ h$. This means that from the topological point of view, $\mathcal{C}^2$ diffeomorphisms and irrational rotations are exactly the same.

A natural question, the so called rigidity question, arises at this point: are diffeomorphisms and irrational rotations the same also from the geometrical point of view? In other words: is the conjugating homeomorphism h a smooth function? What we are really asking is that the orbits of diffeomorphisms are in small scale geometrically the same than the orbits of the rotations. This means that the geometry of diffeomorphisms is rigid: it cannot be modified on asymptotic small scale. This seems a very strong statement and unlikely to be true.

However in the previous chapter we proved that for $\mathcal{C}^3$ diffeomorphisms the fast renormalizations converge to rotations. This result says that in small scale the first return map is asymptotically a rotation independently of the original diffeomorphism. In small scale the orbits of any $\mathcal{C}^3$ diffeomorphism are described by a rotation. This statement and extra conditions on the rotation number will lead to prove rigidity. Precisely: there exists a set $E \subset [0, 1]$ of full measure such that for all $\mathcal{C}^3$ diffeomorphism f with rotation number in E, the conjugating map h is $\mathcal{C}^1$.

The reason for the condition of the rotation number is related to the fact that fast renormalizations converge to rotations but the intermediate ones a priori only converge to affine maps. The condition on the rotation number to be in E implies that all renormalizations converge to rotations. In Chap. 9, we will construct an analytic circle map with Liouvillian rotation number for which the intermediate renormalizations don't converge fst enough to rotations and the conjugacy will not be $\mathcal{C}^1$.

Y. Dong et al., *One-Dimensional Dynamics*, Lecture Notes of the Unione Matematica Italiana 30, https://doi.org/10.1007/978-3-031-94882-4_8

8.1 Rigidity

By Theorem 3.3.1, if f is a C^2 diffeomorphism with irrational rotation number ρ there exists an homeomorphism h such that $h \circ f = r_\rho \circ h$. We say that f is C^1 conjugate to the rotation if h is a C^1 diffeomorphism.

Remark 8.1.1 In Proposition 4.3.3 we constructed a homeomorphism h in a different way than in Theorem 3.3.1. In this chapter we will prove that this h constructed from the renormalization procedure is a C^1 diffeomorphism. This implies that all conjugacies between f and r_ρ are C^1. In fact, suppose $\tilde{h}$ is another conjugacy, then we have $f = h^{-1} \circ r_\rho \circ h = \tilde{h}^{-1} \circ r_\rho \circ \tilde{h}$. Then $H = \tilde{h} \circ h^{-1}$ commutes with r_ρ, which implies that $DH \equiv 1$ and H is of the form $x \mapsto x + \theta$. Hence $\tilde{h} = H \circ h$ is C^1. Note that this is due to properties of the rotations, and in general the existence of a C^1 conjugacy between two systems does not imply that all conjugacies are C^1.

The following lemma says that if f is C^1 conjugate to the rotation then the intermediate renormalizations also converge to rotations. The proof is left as an exercise, see Exercise 8.4.1.

Lemma 8.1.2 *Let $f \in \mathcal{I}^3$. If f is C^1 conjugate to the rotation then*

$$d_1(R^n f, \mathcal{W}) \longrightarrow 0.$$

Conversely, convergence of renormalization is not enough for rigidity. We introduce now the set of rotation numbers for which rigidity holds. For all $\lambda > 1$, we define E_λ as the set of irrational numbers $\rho = [a_1, a_2, \dots] \in (0, 1)$ for which there exist $C > 0$ and $\tilde{\lambda} < \lambda$ such that

$$a_n \leq C\tilde{\lambda}^n.$$

Lemma 8.1.3 *For all $\lambda > 1$ E_λ has full Lebesgue measure.*

The proof of Lemma 8.1.3 is left as an exercise, see Exercise 8.4.2. We can now state the main theorem of this chapter, the rigidity theorem:

Theorem 8.1.4 *There exists $\lambda_0 > 1$ such that for all diffeomorphisms $f \in \mathcal{I}^3$ having rotation number $\rho(f) \in E_{\lambda_0}$, f is C^1 conjugate to the rotation.*

We introduce now a quantity which measures the rate of convergence of intermediate renormalizations. For all $k \in \mathbb{N}$

$$\epsilon_k(f) = \max_{\alpha_k \leq n \leq \alpha_{k+1}} \left\{ \max_{x \in [0,1]} |DR^n f(x) - 1| \right\}. \tag{8.1.1}$$

Lemma 8.1.5 *There exists $\lambda_0 > 1$ such that for all $f \in \mathcal{I}^3$ having rotation number $\rho(f) \in E_{\lambda_0}$, $\sum_k \epsilon_k(f) < \infty$.*

Proof Let $\lambda_0 = \frac{1}{\rho_0}$ where $\rho_0 < 1$ comes from Theorem 7.1.3. In particular, since

$$|DR^{\alpha_k} f - 1| \leq C|\log DR^{\alpha_k} f| \leq Cd_1(R^{\alpha_k} f, \mathcal{W}),$$

there exists a positive constant C_f such that

$$|DR^{\alpha_k} f - 1| \leq C_f \rho_0^k.$$

Let $n = \alpha_k + i$ with $i \leq a_{k+1}$. Then, for every $x \in [0, 1]$,

$$(1 - C_f \rho_0^k)^{a_{k+1}(f)} \leq DR^n f(x) \leq (1 + C_f \rho_0^k)^{a_{k+1}(f)}.$$

The hypothesis $\rho(f) \in E_{\lambda_0}$ implies that

$$a_{k+1}(f) \leq C\tilde{\lambda}^{k+1}$$

with $\tilde{\lambda} < \lambda_0$ and C a positive constant. As consequence

$$\epsilon_k(f) \leq C_f C\tilde{\lambda} \left(\frac{\tilde{\lambda}}{\lambda_0}\right)^k,$$

which implies the lemma. □

Proposition 8.1.6 *Every diffeomorphism $f \in \mathcal{I}^3$ with*

$$\sum_k \epsilon_k(f) < \infty$$

is C^1 conjugate to the rotation.

Observe that Theorem 8.1.4 is a consequence of Proposition 8.1.6 and Lemma 8.1.5. Proposition 8.1.6 is proved in the next section.

8.2 Proof of Rigidity

The aim of this section is to prove Proposition 8.1.6. The proof is divided in the next two subsections.

8.2.1 Scaling Ratios

Let $f \in \mathcal{I}^3$ with $\epsilon_k = \epsilon_k(f) \to 0$. For all $k \in \mathbb{N}$ we denote by L_{α_k} and R_{α_k} the left and right domain of the pre-renormalization corresponding to $R^{\alpha_k} f$. See

Lemma 4.1.6. In the same way we denote by $L^*_{\alpha_k}$ and $R^*_{\alpha_k}$ the corresponding left and right domain of the rotation having the same renormalization type as f. The rest of this subsection on scaling ratios states some properties of these domains.

Lemma 8.2.1 *If $R^{\alpha_k} f \in \mathcal{D}^3_-$, then*

L1 $|L^*_{\alpha_k}| < |R^*_{\alpha_k}|$,

L2 $|L_{\alpha_k}|(1-\epsilon_k) < |R_{\alpha_k}|$,

L3 $\begin{pmatrix} |R^*_{\alpha_k}| \\ |L^*_{\alpha_k}| \end{pmatrix} = \begin{pmatrix} 1 & a_{k+1} \\ 0 & 1 \end{pmatrix} \begin{pmatrix} |R^*_{\alpha_{k+1}}| \\ |L^*_{\alpha_{k+1}}| \end{pmatrix}$,

L4 $\begin{pmatrix} |R_{\alpha_k}| \\ |L_{\alpha_k}| \end{pmatrix} = \begin{pmatrix} 1 & a_{k+1}(1+\tilde{\epsilon}_k) \\ 0 & 1 \end{pmatrix} \begin{pmatrix} |R_{\alpha_{k+1}}| \\ |L_{\alpha_{k+1}}| \end{pmatrix}$ *where* $|\tilde{\epsilon}_k| \le \epsilon_k$.

Proof Let r be the rotation with renormalization type $a(r) = a(f)$ (r has angle the rotation number of f). Observe that $pR^{\alpha_k} r(L^*_{\alpha_k}) \subset R^*_{\alpha_k}$ and that $DpR^{\alpha_k} r = 1$. Recall that pR denotes the pre-renormalization operator, see Sect. 4.1, in particular Lemma 4.1.2. This implies $L1$. $L2$ follows similarly after observing that $|DpR^{\alpha_k} f - 1| \le \epsilon_k$. To prove $L3$ observe that

$$R^*_{\alpha_k} = R^*_{\alpha_{k+1}} \cup \bigcup_{i=1}^{a_{k+1}} (pR^{\alpha_k} r)^i (L^*_{\alpha_{k+1}})$$

where all the parts are pairwise disjoint, and that for all $1 \le i \le a_{k+1}$,

$$(pR^{\alpha_k} r)^i (L^*_{\alpha_{k+1}}) = pR^{\alpha_k + i - 1} r(L^*_{\alpha_{k+1}})$$

and

$$DpR^{\alpha_k + i} r = 1.$$

Similarly for proving $L4$ observe that

$$R_{\alpha_k} = R_{\alpha_{k+1}} \cup \bigcup_{i=1}^{a_{k+1}} (pR^{\alpha_k} f)^i (L_{\alpha_{k+1}})$$

where all the parts are pairwise disjoint, and that for all $1 \le i \le a_{k+1}$,

$$(pR^{\alpha_k} f)^i (L_{\alpha_{k+1}}) = pR^{\alpha_k + i - 1} f(L_{\alpha_{k+1}})$$

and

$$|DpR^{\alpha_k + i} f - 1| \le \epsilon_k.$$

□

Lemma 8.2.2 *If $R^{\alpha_k} f \in \mathcal{D}^3_+$, then*

U1 $|R^*_{\alpha_k}| < |L^*_{\alpha_k}|$,

U2 $|R_{\alpha_k}|(1-\epsilon_k) < |L_{\alpha_k}|$,

U3 $\begin{pmatrix} |R^*_{\alpha_k}| \\ |L^*_{\alpha_k}| \end{pmatrix} = \begin{pmatrix} 1 & 0 \\ a_{k+1} & 1 \end{pmatrix} \begin{pmatrix} |R^*_{\alpha_{k+1}}| \\ |L^*_{\alpha_{k+1}}| \end{pmatrix}$,

U4 $\begin{pmatrix} |R_{\alpha_k}| \\ |L_{\alpha_k}| \end{pmatrix} = \begin{pmatrix} 1 & 0 \\ a_{k+1}(1+\tilde{\epsilon}_k) & 1 \end{pmatrix} \begin{pmatrix} |R_{\alpha_{k+1}}| \\ |L_{\alpha_{k+1}}| \end{pmatrix}$ *where* $|\tilde{\epsilon}_k| \le \epsilon_k$.

Proof The proof is similar to proof of Lemma 8.2.1. □

For all $k \in \mathbb{N}$ we define the scaling ratio θ_k as

$$\theta_k = \frac{|L_{\alpha_k}|}{|R_{\alpha_k}|}$$

and let θ^*_k be the corresponding ratio of the rotation having the same renormalization type as f. In the rest of this subsection we consider the quantity

$$\Delta_k = \frac{\theta_k}{\theta^*_k} - 1$$

and we prove the following proposition.

Proposition 8.2.3 *There exists a positive constant C such that $|\Delta_k| \le C\sum_{\ell \ge k} \epsilon_\ell (\frac{2}{3})^{\ell-k}$.*

The proposition is consequence of the following two lemmas.

Lemma 8.2.4 *There exists a positive constant C such that $|\Delta_k| \le C$ for all k.*

Proof We first consider the case $R^{\alpha_k} f \in \mathcal{D}^3_-$. As consequence $R^{\alpha_{k+1}} f \in \mathcal{D}^3_+$. By property $U1$ of Lemma 8.2.2 we have $\theta^*_{k+1} > 1$ and by $L3$ of Lemma 8.2.1 we get

$$\theta^*_k = \frac{|L^*_{\alpha_{k+1}}|}{|R^*_{\alpha_{k+1}}| + a_{k+1}|L^*_{\alpha_{k+1}}|} = \frac{\theta^*_{k+1}}{1 + a_{k+1}\theta^*_{k+1}} = \frac{1}{a_{k+1} + \frac{1}{\theta^*_{k+1}}}. \tag{8.2.1}$$

As a consequence,

$$\theta^*_k \ge \frac{1}{a_{k+1}+1}. \tag{8.2.2}$$

Similarly by $L4$ of Lemma 8.2.1 we get

$$\theta_k = \frac{\theta_{k+1}}{1 + a_{k+1}(1+\tilde{\epsilon}_k)\theta_{k+1}} \tag{8.2.3}$$

and thus, for k large enough, such that $\epsilon_k \leq 1/2$ (recall that $\epsilon_k \to 0$),

$$a_{k+1}(1+\tilde{\epsilon}_k)\theta_k = \frac{a_{k+1}(1+\tilde{\epsilon}_k)\theta_{k+1}}{1+a_{k+1}(1+\tilde{\epsilon}_k)\theta_{k+1}} \in (0,1)\,.$$

Finally, using (8.2.2),

$$|\Delta_k| = \left|\frac{\theta_k}{\theta_k^*} - 1\right| \leq \frac{\theta_k}{\theta_k^*} + 1 = \frac{a_{k+1}(1+\tilde{\epsilon}_k)\theta_k}{a_{k+1}(1+\tilde{\epsilon}_k)\theta_k^*} + 1 \leq \frac{1}{a_{k+1}(1+\tilde{\epsilon}_k)\cdot\frac{1}{1+a_{k+1}}} + 1 \leq 5$$

for k large enough.

Suppose now that $R^{\alpha_k} f \in \mathcal{D}^3_+$. Then by $U3$ and $U4$ of Lemma 8.2.2 we have

$$\theta_k^* = a_{k+1} + \theta_{k+1}^* > a_{k+1},$$

and

$$\theta_k = a_{k+1}(1+\tilde{\epsilon}_k) + \theta_{k+1}.$$

By $L2$ of Lemma 8.2.1 we have

$$\theta_{k+1} < \frac{1}{1-\epsilon_{k+1}}.$$

Hence, for k large enough such that $\epsilon_k \leq 1/2$,

$$|\Delta_k| \leq \frac{\theta_k}{\theta_k^*} + 1 \leq \frac{a_{k+1}(1+\tilde{\epsilon}_k) + \frac{1}{1-\epsilon_{k+1}}}{a_{k+1}} + 1 \leq 1 + \frac{1}{2} + 2 + 1 \leq 5.$$

The lemma follows. □

Lemma 8.2.5 *For k large enough $|\Delta_k| \leq \frac{2}{3}|\Delta_{k+1}| + 2\epsilon_k$.*

Proof We prove the lemma in the case $R^{\alpha_k} f \in \mathcal{D}^3_-$. The other case is similar. As consequence $R^{\alpha_{k+1}} f \in \mathcal{D}^3_+$. By property $U2$ of Lemma 8.2.2 we have

$$\theta_{k+1} > 1 - \epsilon_{k+1}. \tag{8.2.4}$$

By (8.2.1) and (8.2.3) we get

$$\begin{aligned}|\Delta_k| &= \left|\frac{\theta_{k+1}}{1+a_{k+1}(1+\tilde{\epsilon}_k)\theta_{k+1}}\frac{1+a_{k+1}\theta_{k+1}^*}{\theta_{k+1}^*} - 1\right| \\ &= \left|\frac{\theta_{k+1} - \theta_{k+1}^* - \tilde{\epsilon}_k a_{k+1}\theta_{k+1}\theta_{k+1}^*}{(1+a_{k+1}(1+\tilde{\epsilon}_k)\theta_{k+1})\theta_{k+1}^*}\right|\end{aligned}$$

$$\leq \frac{1}{1+a_{k+1}(1+\tilde{\epsilon}_k)\theta_{k+1}}|\Delta_{k+1}| + \epsilon_k \frac{a_{k+1}\theta_{k+1}}{1+a_{k+1}(1+\tilde{\epsilon}_k)\theta_{k+1}}.$$

By (8.2.4) we get

$$\frac{1}{1+a_{k+1}(1+\tilde{\epsilon}_k)\theta_{k+1}} < \frac{1}{1+a_{k+1}(1+\tilde{\epsilon}_k)(1-\epsilon_{k+1})} \leq \frac{1}{1+1\cdot\frac{1}{\sqrt{2}}\cdot\frac{1}{\sqrt{2}}} = \frac{2}{3}$$

for k large enough such that $\epsilon_k, \epsilon_{k+1} \leq 1 - \frac{1}{\sqrt{2}}$. Moreover, for k large enough,

$$\frac{a_{k+1}\theta_{k+1}}{1+a_{k+1}(1+\tilde{\epsilon}_k)\theta_{k+1}} = \frac{1}{1+\tilde{\epsilon}_k}\cdot\frac{1}{\frac{1}{a_{k+1}(1+\tilde{\epsilon}_k)\theta_{k+1}}+1} \leq 2.$$

This concludes the proof. □

8.2.2 Properties of Dynamical Partitions

Let $f \in \mathcal{I}^3$ with $\epsilon_k(f) \to 0$. Note that the exponential convergence of fast renormalizations to rotations implies that the distortion satisfies

$$\text{Dist}\left(f^{l_{\alpha n}}|_{L_{\alpha n}}\right), \text{Dist}\left(f^{r_{\alpha n}}|_{R_{\alpha n}}\right) \leq 2\max|\log DR^{\alpha_n} f(x)| \leq C\rho^n$$

for some constants $\rho < 1$ and $C > 0$. In this subsection we prove that the distortion is exponentially small for all iterates of f. The precise statement is given in the following proposition.

Proposition 8.2.6 *There exist $\rho < 1$ and $C > 0$ such that*

$$\text{Dist}(f^i|_{L_{\alpha n}}), \text{Dist}(f^j|_{R_{\alpha n}}) \leq C\rho^n$$

where $1 \leq i < l_{\alpha_n}$ and $1 \leq j < r_{\alpha_n}$.

The proof of Proposition 8.2.6 needs some technical lemmas. Note that the proof has similar structure as the proof of Proposition 6.2.7. We will be using Lemma 6.2.6. In particular, observe that Lemma 8.2.7 plays a similar role as (6.2.4). It provides the exponential decay after four iterations.

Lemma 8.2.7 *There exist $\rho < 1$ and $C > 0$ such that for all $\sigma_0 < 1$ there exists $\sigma < 1$ with the following properties.*

Suppose that fast$R^n f \in \mathcal{D}^3_-$, then for all $1 \leq j \leq r_{\alpha_n}$ and for all $x, y \in R_{\alpha_n}$ we have

$$\frac{|x-y|}{|R_{\alpha_n}|} \le \sigma_0 \implies \left|\log \frac{Df^j(x)}{Df^j(y)}\right| \le \sigma \left|\log \frac{Df^j(u_n^+)}{Df^j(c)}\right| + C\rho^n.$$

Suppose that $fastR^n f \in \mathcal{D}^3_+$, *then for all* $1 \le i \le l_{\alpha_n}$ *and for all* $x, y \in L_{\alpha_n}$ *we have*

$$\frac{|x-y|}{|L_{\alpha_n}|} \le \sigma_0 \implies \left|\log \frac{Df^i(x)}{Df^i(y)}\right| \le \sigma \left|\log \frac{Df^i(c)}{Df^i(u_n^-)}\right| + C\rho^n.$$

Proof Assume $fastR^n f \in \mathcal{D}^3_-$. Let $\varphi_k = Nf|_{f^k(R_{\alpha_n})}$ with $0 \le k \le j-1$ and let $m_k : [0,1] \to [0,1]$ be the Möbius map with weight equal to

$$w_k = \int_{f^k(R_{\alpha_n})} \eta_f = \log \frac{Df(f^k(u_n^+))}{Df(f^k(c))}.$$

Then Propositions 5.1.12 and 6.2.7 assure the existence of constants $C > 0$ and $\rho < 1$ such that

$$dist_{\mathcal{C}^1}(O\underline{\varphi}, O\underline{m}) \le C\text{mesh}(\mathscr{P}_{\alpha_n}) \le C\rho^n, \tag{8.2.5}$$

where $\underline{\varphi} = \{\varphi_k\}$ and $\underline{m} = \{m_k\}$. Let $\varphi = O\underline{\varphi}$ and $m = O\underline{m}$. Then, by Lemma 5.1.6, the Möbius diffeomorphism m has weight

$$w = \sum_{k=0}^{j-1} w_k = \log \frac{\prod_{k=0}^{j-1} Df(f^k(u_n^+))}{\prod_{k=0}^{j-1} Df(f^k(c))} = \log \frac{Df^j(u_n^+)}{Df^j(c)}.$$

Note that $|w| \le \int |\eta_f|$ is bounded. Denote the bound by K. Suppose $w < 0$, observe that $\eta_m < 0$ and is strictly increasing on $[0, 1]$. Hence when $|x - y| \le \sigma_0$,

$$\left|\log \frac{Dm(x)}{Dm(y)}\right| = \left|\int_x^y \eta_m\right| \le \left|\int_0^{\sigma_0} \eta_m\right| = \left|\log \frac{Dm(\sigma_0)}{Dm(0)}\right| = |w'|.$$

By Lemma 5.1.5, $e^{-\frac{w'}{2}} - 1 = \sigma_0(e^{-\frac{w}{2}} - 1)$. Suppose $w' = bw$, we show that there exists $\sigma < 1$ such that $b \le \sigma$. Notice that

$$\max_{x\in[w,0]} |(e^{-\frac{x}{2}} - 1)'| = \max_{x\in[w,0]} \frac{1}{2} e^{-\frac{x}{2}} = \frac{1}{2} e^{-\frac{w}{2}},$$

we have

$$(\frac{1}{2} e^{-\frac{w}{2}}) \cdot (1-b)|w| \ge (1-\sigma_0)(e^{-\frac{w}{2}} - 1),$$

and thus

$$1 - b \geq 2(1 - \sigma_0)\frac{1 - e^{\frac{w}{2}}}{|w|}.$$

By the mean value theorem, since $|w| \leq K$ we have

$$\frac{1 - e^{\frac{w}{2}}}{|w|} = \frac{|(1 - e^{\frac{w}{2}}) - 0|}{|w - 0|} \geq \frac{1}{2}e^{-\frac{K}{2}},$$

hence

$$b \leq 1 - (1 - \sigma_0)e^{-\frac{K}{2}}.$$

Similarly, if $w > 0$, then $\eta_m > 0$ and is strictly increasing, we have

$$\left|\log \frac{Dm(x)}{Dm(y)}\right| = \left|\int_x^y \eta_m\right| \leq \left|\int_{1-\sigma_0}^1 \eta_m\right| = \left|\log \frac{Dm(1)}{Dm(1 - \sigma_0)}\right| = |w''|$$

where w'' satisfies $e^{\frac{w''}{2}} - 1 = \sigma_0(e^{\frac{w}{2}} - 1)$, and we can show that $w'' \leq \sigma w$ for some $\sigma < 1$. Hence there exist $\sigma < 1$ such that

$$\left|\log \frac{Dm(x)}{Dm(y)}\right| \leq \sigma\, |w| = \sigma \left|\log \frac{Df^j(u_n^+)}{Df^j(c)}\right| \tag{8.2.6}$$

whenever $|x - y| \leq \sigma_0$. Moreover, using (8.2.5),

$$\left|\log \frac{D\varphi(x)}{Dm(x)}\right| \leq C\rho^n. \tag{8.2.7}$$

From (8.2.6) and (8.2.7) we get, choosing the corresponding rescaled points $x', y' \in [0, 1]$,

$$\begin{aligned}
\left|\log \frac{Df^j(x)}{Df^j(y)}\right| &= \left|\log \frac{D\varphi(x')}{D\varphi(y')}\right| \\
&\leq \left|\log \frac{D\varphi(x')}{Dm(x')}\right| + \left|\log \frac{Dm(y')}{D\varphi(y')}\right| + \left|\log \frac{Dm(x')}{Dm(y')}\right| \\
&\leq \sigma \left|\log \frac{Df^j(u_n^+)}{Df^j(c)}\right| + 2C\rho^n.
\end{aligned}$$

The case when $fastR^n f \in \mathcal{D}_+^3$ can be treated similarly. □

Corollary 8.2.8 *Let $f \in \mathcal{I}^3$. There exist $\rho < 1$ and $C > 0$ such that the following holds.*

Suppose that fast $R^n f \in \mathcal{D}^3_-$, then for all $0 \le j \le r_{\alpha_n}$ we have

$$\left|\mathrm{Dist}(f^j|_{R_{\alpha_n}})\right| \le \left|\log \frac{Df^j(u_n^+)}{Df^j(c)}\right| + C\rho^n.$$

Suppose that fast $R^n f \in \mathcal{D}^3_+$, then for all $0 \le j \le l_{\alpha_n}$ we have

$$\left|\mathrm{Dist}(f^j|_{L_{\alpha_n}})\right| \le \left|\log \frac{Df^j(c)}{Df^j(u_n^-)}\right| + C\rho^n.$$

Proof It is a direct consequence of Lemma 8.2.7 for σ_0 arbitrarily close to one. □

Proof of Proposition 8.2.6 First we introduce some notation. For all $0 \le j \le r_{\alpha_n}$ we define

$$w_{n,j}^+ = \left|\log \frac{Df^j(u_n^+)}{Df^j(c)}\right|$$

and

$$w_n^+ = \max_{0\le j\le r_{\alpha_n}} \left\{w_{n,j}^+\right\}.$$

Similarly, for all $0 \le j \le l_{\alpha_n}$, we define

$$w_{n,i}^- = \left|\log \frac{Df^j(c)}{Df^j(u_n^-)}\right|$$

and

$$w_n^- = \max_{0\le j\le l_{\alpha_n}} \left\{w_{n,j}^-\right\}.$$

Corollary 8.2.8 implies that to prove the proposition it suffices to show that $w_n^{\pm}$ tends to zero exponentially.

We assume that *fast* $R^n f \in \mathcal{D}^3_-$. We obtain an estimate for w_{n+1}^- as follows. Notice that $u_{n+1}^- = u_n^-$. Let $j \le l_{\alpha_{n+1}} = l_{\alpha_n} + a_{n+1} r_{\alpha_n}$. If $j \le l_{\alpha_n}$ then

$$|w_{n+1,j}^-| = \left|\log \frac{Df^j(c)}{Df^j(u_{n+1}^-)}\right| = \left|\log \frac{Df^j(c)}{Df^j(u_n^-)}\right| \le w_n^-. \tag{8.2.8}$$

Let us suppose now that $j = l_{\alpha_n} + k r_{\alpha_n} + t$ for $k < a_{n+1}$ and $t \le r_{\alpha_n}$. From Theorem 7.1.1 it follows that there exist $C > 0$ and $\rho < 1$ such that

$$\mathrm{Dist}\left(f^{l_{\alpha_n}}|_{L_{\alpha_n}}\right), \mathrm{Dist}\left(f^{r_{\alpha_n}}|_{R_{\alpha_n}}\right) \le C\rho^n.$$

This implies that

$$|w_{n+1,j}^{-}| = \left|\log \frac{Df^j(c)}{Df^j(u_n^-)}\right| = \left|\log \frac{Df^{l_{\alpha_n}}(c)\left(\prod_{m=0}^{k-1} Df^{r_{\alpha_n}}(u_{n,m}^+)\right) Df^t(u_{n,k}^+)}{Df^{l_{\alpha_n}}(u_n^-)\left(\prod_{m=1}^{k} Df^{r_{\alpha_n}}(u_{n,m}^+)\right) Df^t(u_{n,k+1}^+)}\right|$$

$$\leq \left|\log \frac{Df^{l_{\alpha_n}}(c)}{Df^{l_{\alpha_n}}(u_n^-)}\right| + \left|\log \frac{Df^{r_{\alpha_n}}(u_n^+)}{Df^{r_{\alpha_n}}(u_{n,k}^+)}\right|$$

$$+ \left|\log \frac{Df^t(u_{n,k}^+)}{Df^t(u_{n,k+1}^+)}\right|$$

$$\leq 2C\rho^n + \left|\log \frac{Df^t(u_{n,k}^+)}{Df^t(u_{n,k+1}^+)}\right|.$$

Now apply Lemmas 6.2.6 and 8.2.7 to obtain $\sigma < 1$ such that the following holds. In the case when $a_{n+1} \neq 1$ or $a_{n+1} = a_{n+2} = 1$ we get

$$\left|\log \frac{Df^j(c)}{Df^j(u_n^-)}\right| \leq 2C\rho^n + \left|\log \frac{Df^t(u_{n,k}^+)}{Df^t(u_{n,k+1}^+)}\right|$$

$$\leq 2C\rho^n + \sigma w_{n,t}^+ + C\rho^n$$

$$\leq \sigma w_n^+ + 3C\rho^n.$$

In the case when $a_{n+1} = 1$ we get $k = 0$ and using Corollary 8.2.8,

$$\left|\log \frac{Df^j(c)}{Df^j(u_n^-)}\right| \leq 2C\rho^n + \left|\log \frac{Df^t(u_n^+)}{Df^t(u_{n+1}^+)}\right|$$

$$\leq 2C\rho^n + w_{n,t}^+ + C\rho^n$$

$$\leq w_n^+ + 3C\rho^n.$$

The next step is to obtain an estimate for w_{n+1}^+. Apply again Lemmas 6.2.6 and 8.2.7 to obtain $\sigma < 1$ such that

$$w_{n+1}^+ \leq \sigma w_n^+ + C\rho^n.$$

Applying the same argument we get similar estimates in the case when *fast* $R^{n+1}f \in \mathcal{D}_+^3$. As a consequence of the above estimates we obtain the following.

- If $a_{n+1} \neq 1$ then

$$w_{n+1}^+ \leq \sigma \cdot w_n^+ + C\rho^n$$

$$w_{n+1}^- \leq \max\{\sigma \cdot w_n^+, w_n^-\} + C\rho^n$$

- If $a_{n+1} = 1$ and $a_{n+2} \neq 1$ then

$$w_{n+1}^{+} \leq \sigma \cdot w_n^{+} + C\rho^n$$
$$w_{n+1}^{-} \leq \max\{w_n^{+}, w_n^{-}\} + C\rho^n$$

- If $a_{n+1} = 1$ and $a_{n+2} = 1$ then

$$w_{n+1}^{+} \leq \sigma \cdot w_n^{+} + C\rho^n$$
$$w_{n+1}^{-} \leq \max\{\sigma \cdot w_n^{+}, w_n^{-}\} + C\rho^n$$

- If $a_{n+2} \neq 1$ then

$$w_{n+2}^{+} \leq \max\{w_{n+1}^{+}, \sigma \cdot w_{n+1}^{-}\} + C\rho^{n+1}$$
$$w_{n+2}^{-} \leq \sigma \cdot w_{n+1}^{-} + C\rho^{n+1}$$

- If $a_{n+2} = 1$ then

$$w_{n+2}^{+} \leq \max\{w_{n+1}^{+}, w_{n+1}^{-}\} + C\rho^{n+1}$$
$$w_{n+2}^{-} \leq \sigma \cdot w_{n+1}^{-} + C\rho^{n+1}$$

This implies, by iterating two steps,

- If $a_{n+2} \neq 1$ then

$$w_{n+2}^{+} \leq \sigma \cdot \max\{w_n^{+}, w_n^{-}\} + 2C\rho^n$$
$$w_{n+2}^{-} \leq \sigma \cdot \max\{w_n^{+}, w_n^{-}\} + 2C\rho^n$$

- If $a_{n+2} = 1$ and $a_{n+1} \neq 1$ then

$$w_{n+2}^{+} \leq \max\{\sigma \cdot w_n^{+}, w_n^{-}\} + 2C\rho^n$$
$$w_{n+2}^{-} \leq \sigma \cdot \max\{w_n^{+}, w_n^{-}\} + 2C\rho^n$$

- If $a_{n+2} = 1$ and $a_{n+1} = 1$ then

$$w_{n+2}^{+} \leq \max\{\sigma \cdot w_n^{+}, w_n^{-}\} + 2C\rho^n$$
$$w_{n+2}^{-} \leq \sigma \cdot \max\{w_n^{+}, w_n^{-}\} + 2C\rho^n.$$

Let $w_n = \max\{w_n^{+}, w_n^{-}\}$. By iterating four steps,

- If $a_{n+2} \neq 1$ or $a_{n+4} \neq 1$ then

$$w_{n+4} \leq \sigma w_n + 4C\rho^n.$$

- If $a_{n+2} = 1$ and $a_{n+4} = 1$ then

$$w_{n+4}^{+} \le \max\{\sigma \cdot w_{n+2}^{+}, w_{n+2}^{-}\} + 2C\rho^{n+2}$$
$$w_{n+4}^{-} \le \sigma \cdot \max\{w_{n+2}^{+}, w_{n+2}^{-}\} + 2C\rho^{n+2}$$

which implies

$$w_{n+4}^{+} \le \sigma \cdot w_n + 4C\rho^n$$
$$w_{n+4}^{-} \le \sigma \cdot w_n + 4C\rho^n$$

Both situations together imply that for every $n \in N$

$$w_{n+4} \le \sigma w_n + 4C\rho^n.$$

This means that w_n decays exponentially. Hence, Proposition 8.2.6 is proved. □

8.2.3 *Smoothness of Conjugacy*

In this subsection we complete the proof of Proposition 8.1.6 showing that the conjugacy is continuously differentiable under the condition $\sum_k \epsilon_k(f) < \infty$. We start with some technical lemmas.

Let $f \in \mathcal{I}^3$ with $\sum_k \epsilon_k(f) < \infty$. We suppose *fast*$R^n f \in \mathcal{D}_{-}^3$. The other case can be treated similarly. Recall that in Sect. 6.2 we introduced the following notation: for $1 \le i \le a_{n+1} + 1$

$$U_{n,i} := \begin{cases} [u_{n,i}^{+}, u_{n,i-1}^{+}] & 1 \le i \le a_{n+1} \\ [c, u_{n+1}^{+}] & i = a_{n+1} + 1 \end{cases}$$

and $U_{n,i,k} := f^k(U_{n,i})$ for $k < r_{\alpha_n}$. Let r be the rotation of the same renormalization type as f. The corresponding objects for the rotation are denoted by $U_{n,i}^{*}$ and $U_{n,i,k}^{*}$.

Lemma 8.2.9 *There exist two constants $C > 0$ and $\rho < 1$ such that*

$$\left| \log \frac{\frac{|U_{n,i,k}^{*}|}{|r^k(R_{\alpha_n}^{*})|}}{\frac{|U_{n,i,k}|}{|f^k(R_{\alpha_n})|}} \right| \le C\rho^n + C \sum_{\ell \ge n} \epsilon_\ell \left(\frac{2}{3}\right)^{\ell - n} \tag{8.2.9}$$

for all $i \le a_{n+1} + 1$ and $k < r_{\alpha_n}$.

Proof By Proposition 8.2.6 there exist $C > 0$ and $\rho < 1$ such that

$$\left|\log \frac{\frac{|U^*_{n,i,k}|}{|r^k(R^*_{\alpha_n})|}}{\frac{|U_{n,i,k}|}{|f^k(R_{\alpha_n})|}}\right| = \left|\log \frac{|U^*_{n,i}|}{|R^*_{\alpha_n}|} - \log \frac{\frac{|U_{n,i,k}|}{|U_{n,i}|}}{\frac{|f^k(R_{\alpha_n})|}{|R_{\alpha_n}|}} - \log \frac{|U_{n,i}|}{|R_{\alpha_n}|}\right| \le \left|\log \frac{\frac{|U^*_{n,i}|}{|R^*_{\alpha_n}|}}{\frac{|U_{n,i}|}{|R_{\alpha_n}|}}\right| + C\rho^n.$$

For $i < a_{n+1} + 1$, by Proposition 8.2.3 and Theorem 7.1.1 there exists constant $C > 0$ such that

$$\left|\log \frac{\frac{|U^*_{n,i}|}{|R^*_{\alpha_n}|}}{\frac{|U_{n,i}|}{|R_{\alpha_n}|}}\right| = \left|\log \frac{\frac{|L^*_{\alpha_n}|}{|R^*_{\alpha_n}|}}{\frac{|(f^{l_{\alpha_n}+ir_{\alpha_n}})(L_{\alpha_n})|}{|R_{\alpha_n}|}}\right| \le \left|\log \frac{\theta^*_n}{\theta_n}\right| + \left|\log \frac{|(f^{l_{\alpha_n}+ir_{\alpha_n}})(L_{\alpha_n})|}{|L_{\alpha_n}|}\right|$$
$$\le C \sum_{\ell \ge n} \epsilon_\ell \left(\frac{2}{3}\right)^{\ell-n} + C\epsilon_n$$

For $i = a_{n+1} + 1$, note that $|L_{\alpha_{n+1}}| = |L_{\alpha_n}|$, $|L^*_{\alpha_{n+1}}| = |L^*_{\alpha_n}|$, by Proposition 8.2.3 there exists $C > 0$ such that

$$\left|\log \frac{\frac{|U^*_{n,i}|}{|R^*_{\alpha_n}|}}{\frac{|U_{n,i}|}{|R_{\alpha_n}|}}\right| = \left|\log \frac{\frac{|R^*_{\alpha_{n+1}}|}{|R^*_{\alpha_n}|}}{\frac{|R_{\alpha_{n+1}}|}{|R_{\alpha_n}|}}\right| = \left|\log \frac{\frac{|L^*_{\alpha_n}|}{|R^*_{\alpha_n}|}}{\frac{|L_{\alpha_n}|}{|R_{\alpha_n}|}} + \log \frac{\frac{|R^*_{\alpha_{n+1}}|}{|L^*_{\alpha_{n+1}}|}}{\frac{|R_{\alpha_{n+1}}|}{|L_{\alpha_{n+1}}|}}\right| \le \left|\log \frac{\theta^*_n}{\theta_n}\right| + \left|\log \frac{\theta_{n+1}}{\theta^*_{n+1}}\right|$$
$$\le C \sum_{\ell \ge n} \epsilon_\ell \left(\frac{2}{3}\right)^{\ell-n} + C \sum_{\ell \ge n+1} \epsilon_\ell \left(\frac{2}{3}\right)^{\ell-n-1}$$
$$\le \left(1 + \left(\frac{2}{3}\right)^{-1}\right) C \sum_{\ell \ge n} \epsilon_\ell \left(\frac{2}{3}\right)^{\ell-n}.$$

Hence there exists constants $C > 0$ and $\rho < 1$ such that (8.2.9) holds. □

In Proposition 4.3.2 we constructed, for every $n \ge 0$, an homeomorphism $h_{\alpha_n} : [0, 1] \to [0, 1]$ with the following properties:

1. $h_{\alpha_n}(\mathscr{P}_{\alpha_n}) = \mathscr{P}^*_{\alpha_n}$,
2. $h_{\alpha_n} : f^i(L_{\alpha_n}) \to r^i(L^*_{\alpha_n})$ is affine and onto for $0 \le i < l_{\alpha_n}$,
3. $h_{\alpha_n} : f^i(R_{\alpha_n}) \to r^i(R^*_{\alpha_n})$ is affine and onto for $0 \le i < r_{\alpha_n}$,
4. if $I \in \mathscr{P}_{\alpha_n}$, then $h_{\alpha_{n+1}}(I) = h_{\alpha_n}(I)$.

Let $I \in \mathscr{P}_{\alpha_n}$ and let $Dh_{\alpha_n}|_I$ be the derivative of h_{α_n} in the interior of I.

Lemma 8.2.10 *There exist two constants $C > 0$ and $\rho < 1$ such that, for all $I \in \mathscr{P}_{\alpha_{n+1}}$,*

$$\left|\log \frac{Dh_{\alpha_{n+1}}|_I}{Dh_{\alpha_n}|_I}\right| \le C\rho^n + C\sum_{\ell\ge n}\epsilon_\ell\left(\frac{2}{3}\right)^{\ell-n}.$$

Proof We assume that *fast*$R^n f \in \mathcal{D}^3_-$. The proof of the other case is similar. Recall that $\mathscr{P}_{\alpha_{n+1}}$ consists of elements in $\mathscr{L}_{\alpha_n}$ and intervals of the form $U_{n,i,k}$, see Remark 6.2.1. Suppose first that $I \in \mathscr{L}_{\alpha_n}$, then $Dh_{\alpha_{n+1}}|_I = Dh_{\alpha_n}|_I$ and the lemma is proved.

Suppose now that $I \notin \mathscr{L}_{\alpha_n}$, then there exist i, k such that $I = U_{n,i,k}$. By Lemma 8.2.9, we have

$$\left|\log \frac{Dh_{\alpha_{n+1}}|_I}{Dh_{\alpha_n}|_I}\right| = \left|\log \frac{\frac{|U^*_{n,i,k}|}{|U_{n,i,k}|}}{\frac{|r^k(R^*_{\alpha_n})|}{|f^k(R_{\alpha_n})|}}\right| = \left|\log \frac{\frac{|U^*_{n,i,k}|}{|r^k(R^*_{\alpha_n})|}}{\frac{|U_{n,i,k}|}{|f^k(R_{\alpha_n})|}}\right| \le C\rho^n + C\sum_{\ell\ge n}\epsilon_\ell\left(\frac{2}{3}\right)^{\ell-n}.$$

□

Observe that every boundary point of an interval in $\mathscr{P}_{\alpha_n}$ is in the orbit of the critical point $O(c)$. So, if $x \notin O(c)$, for every n there exists a unique $I_n \in \mathscr{P}_{\alpha_n}$ such that $x \in I_n$. Define

$$D(x) = \lim_{n\to\infty} Dh_{\alpha_n}(x).$$

Observe that by Lemma 8.2.10 the previous limit exists and in particular,

$$\begin{aligned} |\log D(x)| &= \left|\log Dh_{\alpha_0}(x) + \sum_{n=0}^{\infty}\log\frac{Dh_{\alpha_{n+1}}(x)}{Dh_{\alpha_n}(x)}\right| \\ &\le \sum_{n=0}^{\infty} C\left(\rho^n + \sum_{\ell\ge n}\epsilon_\ell\left(\frac{2}{3}\right)^{\ell-n}\right) \\ &\le C\left(\frac{1}{1-\rho} + \left(\sum_{k=0}^{\infty}\left(\frac{2}{3}\right)^k\right)\sum_{\ell\ge 0}\epsilon_\ell\right) < \infty. \end{aligned}$$

Hence $\inf_x D(x) > 0$.

Lemma 8.2.11 *The function* $D : [0,1]\setminus O(c) \to (0,\infty)$ *is continuous.*

Proof Fix $n_0 \in \mathbb{N}$, then by Lemma 8.2.10

$$\left|\log \frac{D(x)}{Dh_{\alpha_{n_0}}(x)}\right| \le \sum_{k\ge 0}\left|\log\frac{Dh_{\alpha_{n_0+k+1}}(x)}{Dh_{\alpha_{n_0+k}}(x)}\right|$$

$$\leq \sum_{k\geq 0}\left(C\rho^{n_0+k} + C\sum_{\ell\geq n_0+k}\epsilon_\ell\left(\frac{2}{3}\right)^{\ell-(n_0+k)}\right)$$

$$\leq C_1\left(\rho^{n_0} + \sum_{l\geq n_0}\epsilon_l\right).$$

As consequence, if $x, y \in I_{n_0}$ then

$$\left|\log\frac{D(x)}{D(y)}\right| \leq \left|\log\frac{D(x)}{Dh_{\alpha_{n_0}}(x)}\right| + \left|\log\frac{D(y)}{Dh_{\alpha_{n_0}}(y)}\right| \leq 2C_1\left(\rho^{n_0} + \sum_{l\geq n_0}\epsilon_l\right) \tag{8.2.10}$$

which means that D is continuous. □

Corollary 8.2.12 *There exist two constants $C > 0$ and $\rho < 1$ such that, for every $x \notin O(c)$,*

$$|D(x) - Dh_{\alpha_n}(x)| \leq C\rho^n + C\sum_{l\geq n}\epsilon_l.$$

Lemma 8.2.13 *Let $c_k = f^k(c)$ with $k \geq 0$, then*

$$\lim_{x\uparrow c_k} D(x) = \lim_{x\downarrow c_k} D(x).$$

In particular D extends to a positive continuous function $D : [0, 1] \to \mathbb{R}$.

Proof Fix $k \geq 0$. There exists $n_0 \in \mathbb{N}$ such that for all $n \geq n_0$ you can find two intervals $I_-(c_k, n), I_+(c_k, n) \in \mathscr{P}_{\alpha_n}$ with the property

$$c_k \in \partial I_-(c_k, n) \cap \partial I_+(c_k, n).$$

Define

$$D_+(c_k) = \lim_{n\to\infty} Dh_{\alpha_n}|_{I_+(c_k,n)}$$

and

$$D_-(c_k) = \lim_{n\to\infty} Dh_{\alpha_n}|_{I_-(c_k,n)}.$$

Observe that for all n, because $I_+(c_k, n+1) \subset I_+(c_k, n)$ we have

$$Dh_{\alpha_n}|_{I_+(c_k,n)} = Dh_{\alpha_n}|_{I_+(c_k,n+1)}.$$

Hence, for any j,

$$
\begin{aligned}
D_+(c_k) &= \lim_{n\to\infty} Dh_{\alpha_j}|_{I_+(c_k,j)} \cdot \frac{Dh_{\alpha_{j+1}}|_{I_+(c_k,j+1)}}{Dh_{\alpha_j}|_{I_+(c_k,j+1)}} \cdot \dots \cdot \frac{Dh_{\alpha_n}|_{I_+(c_k,n)}}{Dh_{\alpha_{n-1}}|_{I_+(c_k,n)}} \\
&= Dh_{\alpha_j}|_{I_+(c_k,j)} \cdot \prod_{n=j}^{\infty} \frac{Dh_{\alpha_{n+1}}|_{I_+(c_k,n+1)}}{Dh_{\alpha_n}|_{I_+(c_k,n+1)}},
\end{aligned}
$$

which is indeed convergent because of Lemma 8.2.10. Similarly, for $x \in I_+(c_k, j)$,

$$
\begin{aligned}
D(x) &= \lim_{n\to\infty} Dh_{\alpha_j}(x) \cdot \frac{Dh_{\alpha_{j+1}}(x)}{Dh_{\alpha_j}(x)} \cdot \dots \cdot \frac{Dh_{\alpha_n}(x)}{Dh_{\alpha_{n-1}}(x)} \\
&= Dh_{\alpha_j}(x) \cdot \prod_{n=j}^{\infty} \frac{Dh_{\alpha_{n+1}}(x)}{Dh_{\alpha_n}(x)}.
\end{aligned}
$$

Hence, if $x \in I_+(c_k, j)$, by Lemma 8.2.10 we have

$$
\begin{aligned}
\left|\log \frac{D(x)}{D_+(c_k)}\right| &= \left|\sum_{n=j}^{\infty} \log \frac{Dh_{\alpha_{n+1}}(x)}{Dh_{\alpha_n}(x)} - \sum_{n=j}^{\infty} \log \frac{Dh_{\alpha_{n+1}}|_{I_+(c_k,n+1)}}{Dh_{\alpha_n}|_{I_+(c_k,n+1)}}\right| \\
&\le 2C \sum_{n=j}^{\infty} \left(\rho^n + \sum_{\ell \ge n} \epsilon_\ell \left(\frac{2}{3}\right)^{\ell - n}\right) \to 0 \quad \text{as } j \to \infty.
\end{aligned}
\tag{8.2.11}
$$

For all $\varepsilon > 0$, (8.2.11) implies that there exists a J, such that, for all $j \ge J$,

$$
2C \sum_{n=j}^{\infty} \left(\rho^n + \sum_{\ell \ge n} \epsilon_\ell \left(\frac{2}{3}\right)^{\ell - n}\right) < \varepsilon.
$$

Let $\delta = |I_+(c_k, J)|$. In particular, for all $x > c_k$ with $|x - c_k| < \delta$, we have $x \in I_+(c_k, j)$ for some $j \ge J$, and

$$
\left|\log \frac{D(x)}{D_+(c_k)}\right| < \varepsilon.
$$

This implies that

$$
D_+(c_k) = \lim_{x \downarrow c_k} D(x).
$$

Similarly we have

$$
D_-(c_k) = \lim_{x \uparrow c_k} D(x).
$$

By the mean value theorem there exist $\theta_n^+ \in I_+(c, n)$ and $\theta_n^- \in I_-(c, n)$ such that

$$\begin{aligned}\frac{D_+(c_k)}{D_-(c_k)} &= \lim_{n\to\infty} \frac{|I_+^*(c_k^*, n)|}{|I_+(c_k, n)|} \frac{|I_-(c_k, n)|}{|I_-^*(c_k^*, n)|} \\ &= \lim_{n\to\infty} \frac{|I_+^*(c^*, n)|}{Df^k(\theta_n^+)|I_+(c, n)|} \frac{Df^k(\theta_n^-)|I_-(c, n)|}{|I_-^*(c^*, n)|} \\ &= \lim_{n\to\infty} \frac{|I_+^*(c^*, n)|}{|I_+(c, n)|} \frac{|I_-(c, n)|}{|I_-^*(c^*, n)|} = \frac{D_+(c)}{D_-(c)} \\ &= \lim_{n\to\infty} \frac{|R_{\alpha_n}^*|}{|R_{\alpha_n}|} \frac{|L_{\alpha_n}|}{|L_{\alpha_n}^*|} = 1\end{aligned}$$

where we used Proposition 8.2.3. Notice that we also used that f is a circle diffeomorphism, in particular,

$$\lim_{n\to\infty} \frac{Df^k(\theta_n^+)}{Df^k(\theta_n^-)} = 1.$$

Hence, D extends to a continuous function $D : [0, 1] \to \mathbb{R}$. □

Proof of Proposition 8.1.6 By Proposition 4.3.3 the limit of h_{α_n}, h conjugates f and r. We prove now that h is C^1. Let $x \in [0, 1]$, then by Corollary 8.2.12,

$$h(x) = \lim_{n\to\infty} h_{\alpha_n}(x) = \lim \int_0^x Dh_{\alpha_n} = \int_0^x D.$$

As consequence h is differentiable and

$$Dh(x) = D(x).$$

Because D is a continuous and positive function, h is a C^1 diffeomorphism.

8.3 $C^{1+\alpha}$ Rigidity

In this section we refine the notion of rigidity, see Sect. 8.1. For any constant $\alpha > 0$ we say that a circle diffeomorphism f is $C^{1+\alpha}$ conjugate to a rotation, if there exists a conjugacy h between f and the corresponding rotation r_ρ that is $C^{1+\alpha}$ continuous, i.e., h is a C^1 diffeomorphism and the derivative Dh is α-Hölder continuous. We refer to this phenomenon as $C^{1+\alpha}$ rigidity. Note that similar as in Remark 8.1.1, the existence of one $C^{1+\alpha}$ conjugacy h implies that all conjugacies between f and r_ρ are $C^{1+\alpha}$ continuous.

In this section we prove that for maps with bounded combinatorics, namely $f \in \mathcal{I}_A^3$ (see (7.2.1)), the conjugacy h in Proposition 8.1.6 is $C^{1+\alpha}$.

Theorem 8.3.1 *Let $f \in \mathcal{I}_A^3$ for some $A > 0$. Then there exists constant $\alpha > 0$ such that f is $C^{1+\alpha}$ conjugate to rotation.*

The proof of Theorem 8.3.1 is a combination of the following two lemmas. The first one states that the fluctuations of Dh over intervals of the partition $\mathscr{P}_{\alpha_n}$ decay exponentially. The second lemma states that the intervals in the partition $\mathscr{P}_{\alpha_n}$ are at the least exponentially small. This is the reason why Dh is Hölder. For maps with bounded combinatorics the proof is straightforward and clearly shows the main idea: small fluctuations over large intervals.

Indeed, $C^{1+\alpha}$ rigidity also holds in the larger class of maps with rotation number in E_{λ_0}. In this case the proof is more cumbersome and it does not contribute to the understanding of the nature of the problem.

Lemma 8.3.2 *There exist $C > 0$ and $\rho < 1$ such that, for all $x, y \in I$ with $I \in \mathscr{P}_{\alpha_n}$,*

$$\left|\log \frac{D(x)}{D(y)}\right| \leq C\rho^n.$$

Proof Recall that since $f \in \mathcal{I}_A^3$, by Theorem 7.2.1 there exist constants $C > 0$, $\xi \in (0, 1)$ such that, for all $n \in \mathbb{N}$,

$$\max_{x \in [0,1)} |DR^n f(x) - 1| \leq d_1(R^n f, \mathcal{W}) \leq C\xi^n.$$

Hence by the definition of ϵ_ℓ, see (8.1.1), we have $\epsilon_\ell \leq C\xi^\ell$ for all $\ell \in \mathbb{N}$. The lemma follows from (8.2.10). □

Let us now prove the second lemma, which is a direct consequence of Proposition 7.2.3 and the distortion bounds from Lemma 4.6.1. For all $n \in \mathbb{N}$, let $\theta(n)$ denote the size of the smallest interval of the nth dynamical partition $\mathscr{P}_n$, i.e.,

$$\theta(n) = \min\{\, |I| : I \in \mathscr{P}_n \,\}.$$

Lemma 8.3.3 *Let $f \in \mathcal{I}_A^3$. Then there exist constants $C > 0$, $\sigma \in (0, 1)$ such that for all $n \in \mathbb{N}$,*

$$\theta(\alpha_n) \geq C\sigma^n.$$

Proof By Proposition 7.2.3, there exist constants $\sigma_1, \sigma_2 \in (0, 1)$ such that $\sigma_1 \leq \sigma(n) \leq \sigma_2$ for all $n \in \mathbb{N}$. Without loss of generality let us assume that $R^n f \in \mathcal{D}_-^3$ and prove that $\theta(n+1) \geq \sigma\theta(n)$ for some constant σ. Similar as in Sect. 7.2 we denote $J_0 := R_n \setminus R_{n+1}$. Recall that $\mathscr{P}_{n+1}$ consists of $\mathscr{L}_n = \{f^j(L_n) : 0 \leq j < l_n\}$, $\{f^k(J_0) : 0 \leq k < r_n\}$ and $\mathscr{R}_{n+1} = \{f^k(R_{n+1}) : 0 \leq k < r_n\}$. We only need to show that the elements in the last two collections have size at least $\sigma\theta(n)$. By the definition of scaling ratio,

$$|R_{n+1}| \geq \sigma_1 |R_n|, \quad |J_0| \geq (1-\sigma_2)|R_n|. \tag{8.3.1}$$

Note that

$$\frac{|f^k(R_{n+1})|}{|f^k(R_n)|} = \frac{|f^k(R_{n+1})|}{|R_{n+1}|} \cdot \frac{|R_{n+1}|}{|R_n|} \cdot \frac{|R_n|}{|f^k(R_n)|}. \tag{8.3.2}$$

By the distortion bounds in Lemma 4.6.1, there exists a constant K such that

$$\left| \log \frac{\frac{|f^k(R_{n+1})|}{|R_{n+1}|}}{\frac{|f^k(R_n)|}{|R_n|}} \right| \leq \mathrm{Dist}(f^k|_{R_n}) \leq K.$$

Hence there exists a constant $C > 0$ such that

$$\frac{|f^k(R_{n+1})|}{|R_{n+1}|} \cdot \frac{|R_n|}{|f^k(R_n)|} \geq C > 0.$$

Combining this with (8.3.1) and (8.3.2) we get for all $0 \leq k < r_n$,

$$|f^k(R_{n+1})| \geq C\sigma_1 |f^k(R_n)| \geq C\sigma_1\theta(n).$$

Similarly we can prove that for all $0 \leq k < r_n$,

$$|f^k(J_0)| \geq C(1-\sigma_2)|f^k(R_n)| \geq C(1-\sigma_2)\theta(n).$$

Let $\sigma_3 := C \cdot \min\{\sigma_1, 1-\sigma_2\} > 0$, we have proved that $\theta(n+1) \geq \sigma_3\theta(n)$. Therefore

$$\theta(\alpha_n) \geq \sigma_3^{\alpha_n}\theta(0) \geq \sigma_3^{An}\theta(0) = \sigma^n \min\{c_f, 1-c_f\},$$

where we wrote $\sigma := \sigma_3^A$ and used the fact that $\alpha_n \leq An$ for all $n \in \mathbb{N}$. □

Proof of Theorem 8.3.1 Observe that rotation numbers of bounded type are contained in the class E_{λ_0}. By Theorem 8.1.4, the conjugacy h is a C^1 diffeomorphism. In particular we have already proved the continuity of Dh in Proposition 8.1.6, hence it is enough to show that Dh is α-Hölder continuous, i.e., there exist constants $C, \alpha > 0$ such that, for all $x, y \in [0, 1)$,

$$|Dh(x) - Dh(y)| \leq C|x-y|^\alpha.$$

Fix $x, y \in [0, 1)$. Without loss of generality assume $x < y$. Let $n \in \mathbb{N}$ be the smallest integer such that there exists an element $I_n \in \mathscr{P}_{\alpha_n}$ with $I_n \subset [x, y]$. The existence of such an integer n is guaranteed by the fact that $\text{mesh}(\mathscr{P}_n) \to 0$ as $n \to \infty$ (Lemma 4.2.6). By Lemma 8.3.3, there exist constants $C_1, \sigma > 0$ such that

$$|x - y| \geq |I_n| \geq \theta(\alpha_n) \geq C_1 \sigma^n. \tag{8.3.3}$$

We claim that the difference between the derivatives of Dh_{α_n} at the points x and y is exponentially small. Indeed, observe that by our choice of n, $[x, y]$ is contained in either one element of the previous level dynamical partition $\mathscr{P}_{\alpha_{n-1}}$ or two adjacent elements of $\mathscr{P}_{\alpha_{n-1}}$.

Suppose that there exists $I_{n-1} \in \mathscr{P}_{\alpha_{n-1}}$ such that $[x, y] \subset I_{n-1}$. By Lemma 8.3.2, there exist $C > 0$ and $\rho < 1$ such that

$$|Dh(x) - Dh(y)| \leq C\rho^{n-1} \leq 2C\rho^{n-1} \tag{8.3.4}$$

where we used that Dh is continuous and positive, see Lemma 8.2.13.

Suppose now that there exist two adjacent intervals $I_{n-1}, J_{n-1} \in \mathscr{P}_{\alpha_{n-1}}$ such that $[x, y] \subset I_{n-1} \cup J_{n-1}$. We can assume that $x \in I_{n-1}$ and $y \in J_{n-1}$. Denote the common end point of I_{n-1} and J_{n-1} by z. By using again Lemmas 8.3.2 and 8.2.13, there exist $C > 0$ and $\rho < 1$ such that

$$|Dh(x) - Dh(y)| \leq |Dh(x) - Dh(z)| + |Dh(z) - Dh(y)| \leq 2C\rho^{n-1}. \tag{8.3.5}$$

Let $\alpha > 0$ such that $\sigma^\alpha = \rho$ where σ and ρ are given in the previous two lemmas. Then (8.3.4), (8.3.5), and (8.3.3) imply that

$$|Dh(x) - Dh(y)| \leq 2\frac{C}{\rho}\rho^n = 2\frac{C}{\rho}\left(\sigma^n C_1\right)^\alpha \frac{1}{C_1^\alpha} \leq \frac{2C}{\rho C_1^\alpha}|x - y|^\alpha.$$

8.4 Exercises

Exercise 8.4.1 Prove Lemma 8.1.2.

Exercise 8.4.2 The aim of the exercise is to prove Lemma 8.1.3. Define

$$E_{C,\lambda,n} = \left\{\rho = [a_1, a_2, \ldots] \in [0, 1] |\ a_k \leq C(\lambda)^k \ \forall k \leq n\right\}$$

with $C > 0$, $\lambda > 1$, $n \in \mathbb{N}$ and prove the following steps.

1. There exists a constant K such that for every C, λ, n we have

$$|E_{C,\lambda,n+1}| \geq |E_{C,\lambda,n}| \left(1 - \frac{K}{C\lambda^n}\right)$$

   [Hint:show that the branches of the iterates of the Gauss map have uniformly bounded distortion.]
2. For every λ there exists K_λ such that for every C, $|E_{C,\lambda}| \geq 1 - \frac{K_\lambda}{C}$.
3. $|E_\lambda| = 1$.

Chapter 9
Non-rigidity

Circle diffeomorphims with well behaved rotation number are rigid, i.e. the conjugacies to the corresponding rotations are differentiable, Theorem 8.1.4. The exponential convergence of fast renormalization to rotations is the main reason for this rigidity phenomenon, Theorem 7.1.1. This chapter presents two examples of maps which show that the convergence of renormalizations to rotations is a delicate matter.

The first example is such that there is a subsequence of renormalizations which does not converge to the rotations. Indeed, this example is also not smoothly conjugated to the corresponding rotation. An example of non-rigidity.

The second example is an affine circle map with rotation number in E_{λ_0} such that there is a subsequence of renormalizations which does not converge exponentially to the rotations.

9.1 Non Convergence of Renormalization

In this section we prove the existence of analytic circle diffeomorphisms with irrational rotation number whose conjugacies to rotations are not differentiable. We refer to these maps as examples of non-rigidity. In particular, these are examples of maps whose renormalizations do not converge to rotation (see Exercise 9.3.2).

Theorem 9.1.1 *There exist analytic circle diffeomorphisms with irrational rotation number whose conjugacies to the rotations are not differentiable. In particular, the conjugacies are not differentiable in a dense set of points.*

The examples of non-rigidity given here are inspired by the ones constructed by Arnol'd. Indeed, he gave examples of analytic circle diffeomorphisms whose invariant measures are singular with respect to the Lebesgue measure. This implies

Y. Dong et al., *One-Dimensional Dynamics*, Lecture Notes of the Unione Matematica Italiana 30, https://doi.org/10.1007/978-3-031-94882-4_9

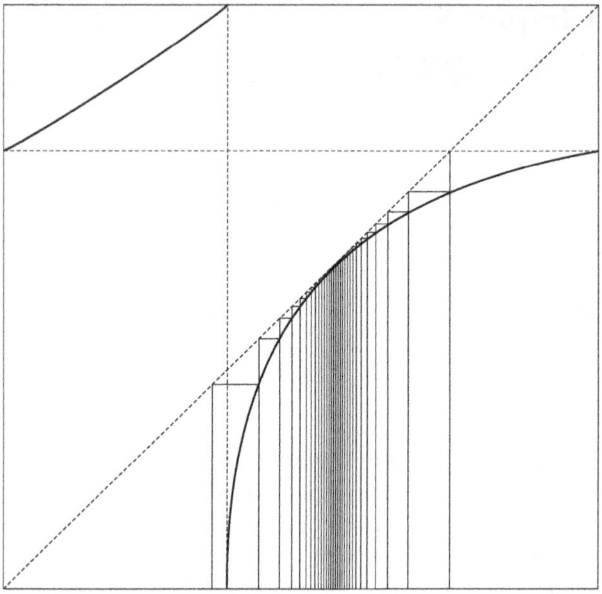

Fig. 9.1 Concentration near Saddle, $a = 29$

non-rigidity. Indeed, if the conjugacy h would be everywhere differentiable then the invariant measure would be given by

$$\mu(I) = |h^{-1}(I)|$$

which is absolutely continuous with density $1/Dh$. Our simplified construction avoids the description of the invariant measures. We only concentrate on creating a point where the conjugacy is not differentiable.

Our non-rigidity examples, as well as the ones by Arnol'd, are perturbations of maps having a periodic point with derivative one. These periodic orbits are called saddle nodes. A peculiarity of our maps is that the orbit of $u^+_{n,i}$, for $i = 0, \dots, a_n$ of the nth fast renormalization (see Lemma 6.2.6) is concentrated near the saddle node, See Fig. 9.1. On the other hand, the corresponding orbit of the corresponding fast renormalization of the corresponding irrational rotation is uniformly distributed, see Fig. 9.2. This essential geometrical difference is responsible for the non-differentiability of the conjugacy.

The saddle node effect, described in the previous paragraph, is especially noticeable and responsible for the non-differentiability of the conjugacy when the irrational rotation number is very close to rational numbers, i.e. the numbers a_n are very large. Compare with Theorem 8.1.4 which shows that if the sequence a_n grows at most exponentially with a bounded rate, the conjugacy is continuously differentiable. The exact characterization of rotation numbers ensuring rigidity is given by Yoccoz in [Yoc95].

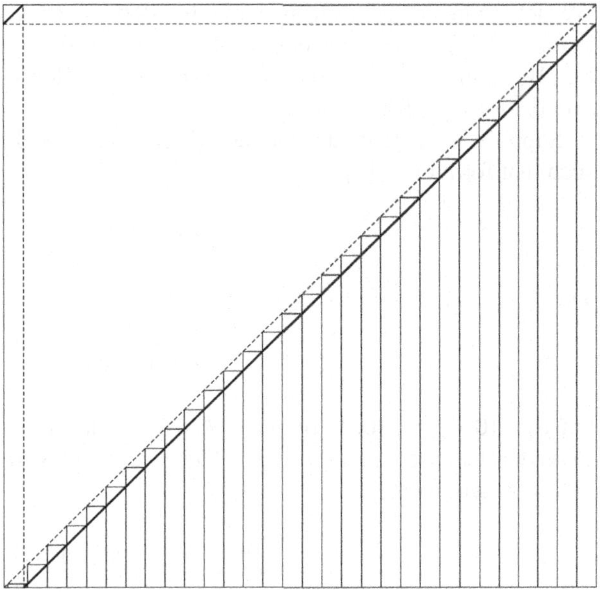

Fig. 9.2 Uniform distribution, $a = 29$

The rest of this chapter is devoted to the proof of Theorem 9.1.1. Consider the family $r_v : [0, 1) \to [0, 1)$ of rotations with $r_v(0) = v \in [0, 1)$. The proof of the following lemma is left as Exercise 9.3.1.

Lemma 9.1.2 *For every finite sequence* $a_1, \dots, a_n$ *with* $a_1 \geq 0$ *and* $a_i \geq 1$ *for* $2 \leq i \leq n$*, there exists a unique open interval* $W_n^* = (v_n^-, v_n)$*,* $v_n^-, v_n \in \mathbb{Q}$ *such that the map*

$$W_n^* \ni v \mapsto \rho(fastR^n(r_v)) \in (0, 1),$$

is an increasing diffeomorphism. In particular, the interval W_n^* *consists exactly of the rotations of renormalization type* $a_1, \dots, a_n$.

Given a finite sequence $a_1, \dots, a_n$, let

$$u_n^* = \max_{v \in W_n^*} \left| U_{\alpha_n}^*(v) \right|,$$

where $U_{\alpha_n}^*(v)$ is the renormalization domain of the nth fast renormalizations of r_v, namely $U_{\alpha_n}^*(v) = L_{\alpha_n}^*(v) \cup R_{\alpha_n}^*(v)$.

Let f be an analytic circle diffeomorphism with $f(0) = 0$. Consider the family of circle diffeomorphisms represented by

$$f_t = r_t \circ f, \ t \in [0, 1).$$

We are going to perform an inductive construction. Let $a_1 = 1$ and W_1^* as in Lemma 9.1.2. There exists a unique interval $W_1 = (t_1^-, t_1)$, such that, for all $t \in W_1$, $\rho(f_t) \in W_1^*$ and $\rho(f_{t_1}) = v_1$. This follows by applying Theorem 2.4.1, Propositions 2.4.5 and 2.4.7.

Suppose we defined an interval $W_n = (t_n^-, t_n)$ and a finite sequence $a_1, \ldots, a_n$, such that for all $t \in W_n$,

$$\rho(f_t) \in W_n^*,$$

and

$$\rho(f_{t_n}) = v_n.$$

Let $g_n : [0, 1) \to [0, 1)$ be $g_n = fastR^n f_{t_n}$ and let c_n be the cutting point of g_n. Consider the case when n is even. The case n odd is treated similarly. Observe that, by using Lemma 9.1.2,

$$\lim_{t \to t_n, t < t_n} \rho(fastR^n f_t) = 1.$$

This implies that

$$\lim_{t \to t_n, t < t_n} a_{n+1}(f_t) = \infty.$$

In particular there exists a saddle node $b_n \in (c_n, 1)$, i.e. $g_n(b_n) = b_n$ and $g_n'(b_n) = 1$. Observe that there might be multiple saddle nodes. We choose the left most. In particular we may assume that $g_n(x) < x$ for all $x \in (c_n, b_n)$, see Fig. 9.3.

Let

$$\beta_n = \frac{b_n - c_n}{1 - c_n} > 0,$$

$$\gamma_n = \min_{t \in W_n} |R_{\alpha_n}(f_t)|,$$

and for $t \in W_n$ let $c_n(t)$ be the cutting point of $fastR^n f_t$. By shrinking W_n, i.e. take t_n^- close enough to t_n we can assure that γ_n is positive.

To describe the accumulation at the saddle node, consider the points

$$z_k^n = g_n^{-k}(c_n) \in (c_n, 1),$$

with $k > 0$, see Fig. 9.3. Observe that $z_k^n \to b_n$ when $k \to \infty$. Choose $k_n > 0$, such that

$$\left|c_n - z_{k_n}^n\right| \geq \frac{3}{4}\beta_n(1 - c_n). \tag{9.1.1}$$

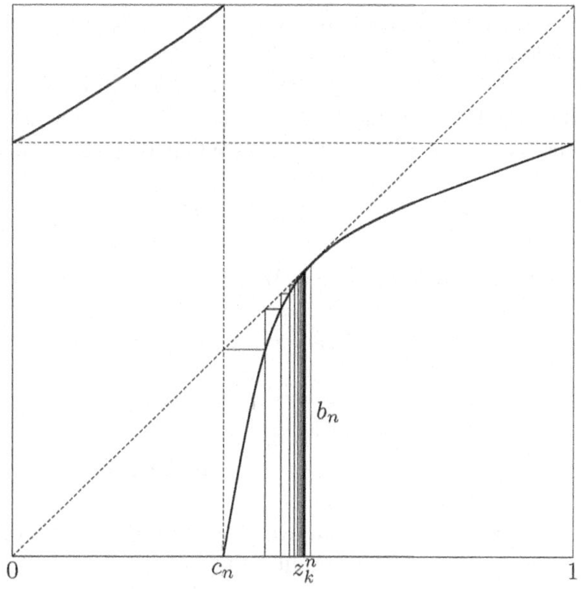

Fig. 9.3 The sequence z_k^n

Since the family is continuous, by shrinking W_n, i.e. take t_n^- close enough to t_n, we may assume that for all $t \in W_n$,

$$z_{k_n}^n(t) = \left(fastR^n f_t\right)^{-k_n}(c_n(t))$$

is well defined and

$$\left|c_n(t) - z_{k_n}^n(t)\right| \geq \frac{2}{3}\beta_n(1 - c_n(t)).$$

Choose A_{n+1} such that,

$$\frac{k_n}{A_{n+1}} \leq \frac{1}{n}\beta_n\frac{\gamma_n}{u_n^*}. \tag{9.1.2}$$

By shrinking W_n, i.e. take $t_n^- < t_n$ even closer, we have that for all $t \in W_n$,

$$a_{n+1}(f_t) \geq A_{n+1}.$$

Let

$$a_{n+1} = \min_{t\in W_n}\{a_{n+1}(f_t)\} + 1.$$

Consider the finite sequence $a_1, \ldots, a_n, a_{n+1}$ and let $W^*_{n+1} = (v^-_{n+1}, v_{n+1})$ be the interval given by Lemma 9.1.2.

By Theorem 2.4.1, Propositions 2.4.5 and 2.4.7 there exists a unique maximal interval W such that $\overline{\{\rho(f_t)\}_{t\in W}} = W^*_{n+1}$. Because $a_{n+1} > \min_{t\in W_n}\{a_{n+1}(f_t)\}$ and by Proposition 2.4.5, we have that $W \subset W_n$. Choose an interval

$$W_{n+1} = (t^-_{n+1}, t_{n+1}) \subset W \subset W_n$$

such that $\{\rho(f_t)\}_{t\in W_{n+1}} = W^*_{n+1}$,

$$\rho(f_{t_{n+1}}) = v_{n+1},$$

and for all $t \in W_{n+1}$,

$$\rho(f_t) \in W^*_{n+1}.$$

This concludes the inductive procedure defining the nested sequence of parameter intervals W_n and the infinite sequence $a_1, a_2, \ldots$. Note that since by our construction $\overline{W_{n+1}} \subset W_n$, the sequence of nested intervals has a non-empty intersection. Let

$$\{t_0\} = \bigcap W_n.$$

Observe that the rotation number $v_0 = \rho(f_{t_0})$ is determined by the infinite sequence $a_1, a_2, \ldots$, see Theorem 4.4.2, and that $v_0 \notin \mathbb{Q}$. In particular, because of Proposition 2.4.7, there is a unique such a t_0.

Let h be the conjugacy between r_{v_0} and f_{t_0} with $h(c^*) = c$ where c^* is the cutting point of r_{v_0} and c is the cutting point of f_{t_0}. The following lemma confirms Theorem 9.1.1.

Lemma 9.1.3 *The conjugacy h is not differentiable at c^*. In particular, h is not differentiable in a dense set, namely at least the orbit of c^*.*

Proof Let $R_{\alpha_n}(f_{t_0})$ and $L_{\alpha_n}(f_{t_0})$ form the renormalization domain of $fastR^n f_{t_0}$. For n even, let $y^n_{k_n} \in R_{\alpha_n}(f_{t_0})$ be the point corresponding to $z^n_{k_n}(t_0)$. Moreover let $x^n_{k_n} \in R^*_{\alpha_n}(v_0)$ be such that $h(x^n_{k_n}) = y^n_{k_n}$. Observe that, if h is differentiable at c^*, then, using (9.1.1) and (9.1.2),

$$Dh(c^*) = \lim_n \frac{|y^n_{k_n} - c|}{|x^n_{k_n} - c^*|} \geq \lim_n \frac{\frac{2}{3}\beta_n R_{\alpha_n}(f_{t_0})}{\frac{k_n}{a_{n+1}}|U^*_n(v_0)|}$$

$$\geq \lim_n \frac{\frac{2}{3}\beta_n \gamma_n}{\frac{k_n}{a_{n+1}}u^*_n} \geq \lim_n \frac{\frac{2}{3}\beta_n \gamma_n}{\frac{k_n}{A_{n+1}}u^*_n} \geq \lim_n \frac{2}{3}n = \infty.$$

This contradicts the assumption that h is differentiable at c^*. □

9.2 Non Exponential Covergence of Renormalization

We have seen in Theorem 7.1.1 that, for a $\mathcal{C}^3$ generalized circle diffeomorphism f, the fast renormalizations $fastR^n f$ converge to the set of rotations $\mathcal{W}$ exponentially, whereas without further assumptions the renormalizations $R^n f$ only converge to the set of affine maps $\mathcal{A}$. In fact, the renormalizations $R^n f$ do not converge to the set of rotations $\mathcal{W}$ for the non-rigidity example presented in the previous section (Exercise 9.3.2). However, recall that if we restrict the rotation number to the rigidity class E_{λ_0}, then the renormalizations $R^n f$ indeed converge to the set of rotations $\mathcal{W}$, see Theorem 8.1.4 and Lemma 8.1.2. Moreover, recall that in Sect. 7.2 we have proved the exponential convergence of $R^n f$ to $\mathcal{W}$ for the maps with bounded combinatorics. A natural question arises: is the convergence of R exponential for all rotation numbers in E_{λ_0}? In this section we provide an example, for which we have no exponential convergence of renormalizations $R^n f$ to the set of rotations $\mathcal{W}$.

Theorem 9.2.1 *There exist affine circle maps $f \in \mathcal{A}$ with rotation number $\rho(f) \in E_{\lambda_0}$, such that the convergence of renormalizations $R^n f$ to the set of rotations $\mathcal{W}$ is not exponential.*

Proof Let $\tilde{\lambda} \in (1, \lambda_0)$ be a fixed constant and let m be a large enough integer such that $\tilde{\lambda}^m > 2$. For simplicity of notations, given a sequence $\{n_j\}_{j\geq 1}$ let us denote by $N_k := \sum_{j=1}^{k} n_j$, $\forall k$. We consider an affine circle map f, not the rotation, with the following renormalization type:

$$\underbrace{1, 1, \ldots, 1}_{n_1 - 1}, a_{N_1}, \underbrace{1, 1, \ldots, 1}_{n_2 - 1}, a_{N_2}, \ldots, \underbrace{1, 1, \ldots, 1}_{n_k - 1}, a_{N_k}, \ldots,$$

where for all k, $n_k := 2m \cdot k$ and $a_{N_k} := \lfloor \tilde{\lambda}^{n_k} \rfloor - 2$. Observe that with this construction

$$\begin{aligned} \alpha_{N_k} &= \sum_{i=1}^{N_k} a_i = N_k - k + \lfloor \tilde{\lambda}^{2m} \rfloor + \cdots + \lfloor \tilde{\lambda}^{2m \cdot k} \rfloor - 2k \\ &= mk^2 + (m-3)k + \lfloor \tilde{\lambda}^{2m} \rfloor + \cdots + \lfloor \tilde{\lambda}^{2m \cdot k} \rfloor \end{aligned} \tag{9.2.1}$$

is exponential in k, and all the long steps a_{N_k} happen in the half space $\mathcal{D}^3_-$. Moreover, the rotation number is in E_{λ_0}. In particular, the renormalizations $R^n f$ converge to the set of rotations by Lemma 8.1.2. We will show that for this affine circle map f,

$$\lim_{k \to \infty} \frac{1}{\alpha_{N_k}} \log \frac{d_1(R^{\alpha_{N_k}} f, \mathcal{W})}{d_1(f, \mathcal{W})} = 0, \tag{9.2.2}$$

i.e., the convergence is not exponential.

Notice that for an affine map f,

$$d_1(R^n f, \mathcal{W}) = \min_{t\in[0,1)} \left(\max\{|c_n - (1-t)|,\ |v_n - t|\}\right) = \frac{1}{2}|1 - c_n - v_n|.$$

Hence

$$d_1(R^n f, \mathcal{W}) = \frac{1}{2}(1-c_n)\left|1 - \frac{v_n}{1-c_n}\right| = \frac{1}{2}(1-c_n)\left|e^0 - e^{s_+(n)}\right|.$$

Recall that $|s_+(n)|, |s_-(n)| \to 0$ as $n \to \infty$, see Proposition 5.3.1. In particular, $|s_+(n)|$ and $|s_-(n)|$ are bounded. Hence there exist constants $C_1, K_1 > 0$ such that

$$C_1(1-c_n)|s_+(n)| \le d_1(R^n f, \mathcal{W}) \le K_1|s_+(n)| \le K_1|s(n)|. \tag{9.2.3}$$

Similarly, there exists a constant $C_2 > 0$ such that

$$d_1(R^n f, \mathcal{W}) = \frac{1}{2}c_n\left|\frac{1-v_n}{c_n} - 1\right| = \frac{1}{2}c_n\left|e^{s_-(n)} - e^0\right| \ge C_2 c_n|s_-(n)|.$$

This and the left inequality of (9.2.3) yield that there exists a constant $K_2 > 0$ such that

$$d_1(R^n f, \mathcal{W}) \ge K_2 \min\{|s_+(n)|,\ |s_-(n)|\},$$

in particular,

$$d_1(R^n f, \mathcal{W}) \ge \begin{cases} K_2|s_+(n)| & \text{if } R^n f \in \mathcal{D}^3_-, \\ K_2|s_-(n)| & \text{if } R^n f \in \mathcal{D}^3_+. \end{cases} \tag{9.2.4}$$

Let us estimate the contraction rate of the d_1 distance after one fast step in the following three different cases.

(i) $a_{n+1} = 1$ and $a_{n+2} = 1$. Suppose $R^{\alpha_n} f \in \mathcal{D}^3_+$. Since $R^{\alpha_{n+1}} f \in \mathcal{D}^3_-$, we have

$$|s_+(\alpha_{n+1})| < |s_-(\alpha_{n+1})|.$$

Since $a_{n+1} = 1$, we have $|s_-(\alpha_{n+1})| = |s_-(\alpha_n)|$ and $|s_+(\alpha_{n+1})| = |s_+(\alpha_n) + s_-(\alpha_n)| = |s_+(\alpha_n)| - |s_-(\alpha_n)|$, we have

$$|s_+(\alpha_n)| < 2|s_-(\alpha_n)|. \tag{9.2.5}$$

Similarly, since $a_{n+2} = 1$ we have

$$|s_+(\alpha_{n+2})| = |s_+(\alpha_{n+1})| = |s_+(\alpha_n)| - |s_-(\alpha_n)|$$

and

$$|s_-(\alpha_{n+2})| = |s_-(\alpha_{n+1})| - |s_+(\alpha_{n+1})| = |s_-(\alpha_n)| - (|s_+(\alpha_n)| - |s_-(\alpha_n)|).$$

Using $|s_+(\alpha_{n+2})| > |s_-(\alpha_{n+2})|$, we have

$$|s_+(\alpha_n)| > \frac{3}{2}|s_-(\alpha_n)|. \tag{9.2.6}$$

By (9.2.3) and (9.2.4), we have

$$\frac{d_1(R^{\alpha_{n+1}} f, \mathcal{W})}{d_1(R^{\alpha_n} f, \mathcal{W})} \geq \frac{K_2|s_+(\alpha_{n+1})|}{K_1|s(\alpha_n)|} = \frac{K_2(|s_+(\alpha_n)| - |s_-(\alpha_n)|)}{K_1(|s_+(\alpha_n)| + |s_-(\alpha_n)|)}.$$

Using (9.2.5) and (9.2.6) and we get in this case

$$\frac{d_1(R^{\alpha_{n+1}} f, \mathcal{W})}{d_1(R^{\alpha_n} f, \mathcal{W})} > \frac{K_2 \cdot \frac{1}{2}|s_-(\alpha_n)|}{K_1 \cdot 3|s_-(\alpha_n)|} = \frac{1}{6}\frac{K_2}{K_1}. \tag{9.2.7}$$

With a similar argument one can show that the estimate (9.2.7) also holds for $R^{\alpha_n} f \in \mathcal{D}^3_-$.

(ii) $a_{n+1} = 1$ and $a_{n+2} = a_{N_j}$ for some j. Namely a_{n+1} is the last 1 before a long step a_{N_j}. Observe that by our construction in this case $R^{\alpha_n} f \in \mathcal{D}^3_+$. Since $a_{n+1} = 1$, similar as in (9.2.5) we have

$$|s_+(\alpha_n)| < 2|s_-(\alpha_n)|.$$

For the other direction observe that $R^{\alpha_{n+2}} f \in \mathcal{D}^3_+$ after a_{N_j} steps in $\mathcal{D}^3_-$, hence

$$|s_+(\alpha_{n+2})| > |s_-(\alpha_{n+2})|,$$

$$|s_+(\alpha_{n+2})| = |s_+(\alpha_{n+1})| = |s_+(\alpha_n)| - |s_-(\alpha_n)|,$$

and

$$|s_-(\alpha_{n+2})| = |s_-(\alpha_{n+1})| - a_{N_j}|s_+(\alpha_{n+1})| = |s_-(\alpha_n)| - a_{N_j}(|s_+(\alpha_n)| - |s_-(\alpha_n)|).$$

As a result

$$|s_+(\alpha_n)| > \frac{a_{N_j} + 2}{a_{N_j} + 1}|s_-(\alpha_n)|.$$

Again using (9.2.3) and (9.2.4) we have

$$\begin{aligned}\frac{d_1(R^{\alpha_{n+1}}f,\mathcal{W})}{d_1(R^{\alpha_n}f,\mathcal{W})} &\geq \frac{K_2(|s_+(\alpha_n)|-|s_-(\alpha_n)|)}{K_1(|s_+(\alpha_n)|+|s_-(\alpha_n)|)}\\ &> \frac{K_2\cdot\frac{1}{a_{N_j}+1}|s_-(\alpha_n)|}{K_1\cdot 3|s_-(\alpha_n)|}\\ &= \frac{1}{3(a_{N_j}+1)}\frac{K_2}{K_1}.\end{aligned} \tag{9.2.8}$$

(iii) $a_{n+1}=a_{N_j}$ for some j. In particular $R^{\alpha_n}f\in\mathcal{D}^3_-$ and $a_{n+2}=1$. Hence

$$|s_+(\alpha_{n+1})| = |s_+(\alpha_n)|,$$

$$|s_-(\alpha_{n+1})| = |s_-(\alpha_n)| - a_{N_j}|s_+(\alpha_n)|,$$

and

$$|s_-(\alpha_{n+1})| < |s_+(\alpha_{n+1})|.$$

As a consequence,

$$|s_-(\alpha_n)| < (a_{N_j}+1)|s_+(\alpha_n)|. \tag{9.2.9}$$

Since $a_{n+2}=1$ we have

$$|s_+(\alpha_{n+2})| = |s_+(\alpha_{n+1})|-|s_-(\alpha_{n+1})| = |s_+(\alpha_n)|-(|s_-(\alpha_n)|-a_{N_j}|s_+(\alpha_n)|),$$

and

$$|s_-(\alpha_{n+2})| = |s_-(\alpha_{n+1})| = |s_-(\alpha_n)| - a_{N_j}|s_+(\alpha_n)|.$$

Using $|s_-(\alpha_{n+2})| > |s_+(\alpha_{n+2})|$ we have

$$|s_-(\alpha_n)| > \left(a_{N_j}+\frac{1}{2}\right)|s_+(\alpha_n)|. \tag{9.2.10}$$

By (9.2.3) and (9.2.4), in this case

$$\frac{d_1(R^{\alpha_{n+1}}f,\mathcal{W})}{d_1(R^{\alpha_n}f,\mathcal{W})} \geq \frac{K_2|s_-(\alpha_{n+1})|}{K_1|s(\alpha_n)|} = \frac{K_2(|s_-(\alpha_n)|-a_{N_j}|s_+(\alpha_n)|)}{K_2(|s_+(\alpha_n)|+|s_-(\alpha_n)|}.$$

Using (9.2.9) and (9.2.10) we get

$$\frac{d_1(R^{\alpha_{n+1}}f,\mathcal{W})}{d_1(R^{\alpha_n}f,\mathcal{W})} > \frac{K_2\cdot\frac{1}{2}|s_+(\alpha_n)|}{K_1\cdot(a_{N_j}+2)|s_+(\alpha_n)|} = \frac{1}{2(a_{N_j}+2)}\frac{K_2}{K_1}. \tag{9.2.11}$$

We are now ready to compute the limit in (9.2.2). Using (9.2.7), (9.2.8), and (9.2.11), we have

$$
\begin{aligned}
\frac{d_1(R^{\alpha_{N_k}} f, \mathcal{W})}{d_1(f, \mathcal{W})} &= \prod_{i=0}^{N_k-1} \frac{d_1(R^{\alpha_{i+1}} f, \mathcal{W})}{d_1(R^{\alpha_i} f, \mathcal{W})} \\
&> \left(\frac{K_2}{K_1}\right)^{N_k} \cdot \prod_{j=1}^{k} \left(\frac{1}{6}\right)^{n_j-2} \cdot \frac{1}{3(a_{N_j}+1)} \cdot \frac{1}{2(a_{N_j}+2)} \\
&= \left(\frac{K_2}{K_1}\right)^{N_k} \cdot \left(\frac{1}{6}\right)^{N_k-k} \cdot \prod_{j=1}^{k} \frac{1}{\lfloor \tilde{\lambda}^{n_j} \rfloor - 1} \cdot \frac{1}{\lfloor \tilde{\lambda}^{n_j} \rfloor} \\
&\geq \left(\frac{K_2}{K_1}\right)^{N_k} \cdot \left(\frac{1}{6}\right)^{N_k} \cdot \prod_{j=1}^{k} \frac{1}{\tilde{\lambda}^{n_j}} \cdot \frac{1}{\tilde{\lambda}^{n_j}} = \left(\frac{K_2}{6K_1 \cdot \tilde{\lambda}^2}\right)^{N_k}.
\end{aligned}
$$

Hence

$$
0 \geq \lim_{k\to\infty} \frac{1}{\alpha_{N_k}} \log \frac{d_1(R^{\alpha_{N_k}} f, \mathcal{W})}{d_1(f, \mathcal{W})} \geq \log \frac{K_2}{6K_1 \cdot \tilde{\lambda}^2} \cdot \lim_{k\to\infty} \frac{N_k}{\alpha_{N_k}} = 0,
$$

where we used that N_k is polynomial and α_{N_k} is exponential (see (9.2.1)). □

9.3 Exercises

Exercise 9.3.1 Prove Lemma 9.1.2. Hint: Observe that the interval W_n^* is an appropriate preimage under the renormalization of rigid rotations, see Lemma 4.5.1.

Exercise 9.3.2 Show that the renormlizations $R^n f$ of the example given in the proof of Theorem 9.1.1 do not converge to $\mathcal{W}$.

Exercise 9.3.3 Given an analytic circle diffeomorphism f with the rotation number constructed in Theorem 9.2.1, is the convergence $d_1(R^n f, \mathcal{W}) \to 0$ exponential?

Exercise 9.3.4 Observe that the set of rotation numbers as selected in Theorem 9.2.1 has measure zero. Is it possible to enlarge this set to a set with positive Lebesgue measure? Namely, does there exist a positive Lebesgue measure set of rotation numbers whose corresponding affine maps do not converge exponentially to rotations?

Chapter 10
The Renormalization Horseshoe

The simplest non trivial dynamical systems are the hyperbolic systems. The prototypical example is Smale's Horseshoe, see Fig. 1.3. A hyperbolic dynamical system is a differentiable system characterized by specific infinitesimal properties at each point. Namely, the tangent space at each point can be split into two parts, the stable and the unstable subspaces. The iterates of the derivatives along the unstable subspace expand exponentially fast while, the iterations of the derivatives along the stable subspace contract exponentially fast. This infinitesimal information of the map has global topological consequences. The starting point to describe the global topological structure of a hyperbolic system is the Stable Manifold Theorem. This theorem assures that for each point x, the so-called stable set $W^s(x)$ consisting of the points y such that

$$d\left(f^n(y), f^n(x)\right) \to 0$$

exponentially fast is (locally) a manifold. Similarly, the unstable set $W^u(x)$ consisting of points y, such that

$$d\left(f^{-n}(y), f^{-n}(x)\right) \to 0$$

exponentially fast is (locally) a manifold. These stable manifolds together form the stable lamination, while the unstable manifolds form the unstable one. For example in Smale's Horseshoe, see Fig. 1.3, the vertical line segments are part of the stable lamination while the horizontal line segments are part of the unstable lamination. There is an extensive literature on hyperbolic systems. Among general references are [BS02, GH83, KH95, Mn87, PM82, PT93, Shu87, Sma65, Sma80].

Observe that the fast renormalization operator defines a dynamical system. In Sect. 10.1 we prove that this dynamical system is not differentiable. Although, because of this lack of differentiability, we cannot say that fast renormalization is hyperbolic, in this chapter we show that it behaves as if it were hyperbolic. Namely,

Y. Dong et al., *One-Dimensional Dynamics*, Lecture Notes of the Unione Matematica Italiana 30, https://doi.org/10.1007/978-3-031-94882-4_10

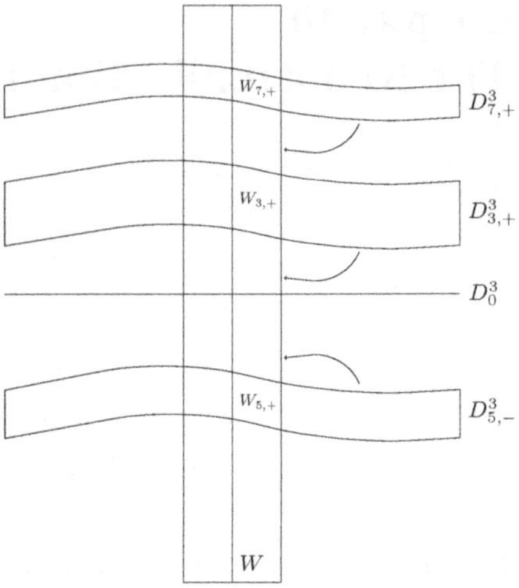

Fig. 10.1 Renormalization horseshoe

there is a stable lamination and only one unstable leaf, see Fig. 10.1. This is called topological hyperbolicity. Furthermore, each leaf of the stable lamination consists exactly of the maps with the same rotation number, while the unstable leaf is just the set of rotations.

Topological hyperbolicity is the main pillar of the theory.

10.1 Non-differentiability of Renormalization

Renormalization has two aspects, restriction and composition. Indeed both are responsible for the non-differentiability of renormalization. Lemma 10.1.1 explains how composition causes non-differentiability of renormalization and Lemma 10.1.2 shows how restriction causes non-differentiability. Unstable and stable is exponential.

Observe that the space $\text{Diff}^3_+([0,1])$ can be identified with $\mathcal{C}^1([0,1])$ by the non-linearity map. Indeed,

$$\eta : \text{Diff}^3_+([0,1]) \to \mathcal{C}^1([0,1]), \quad \varphi \mapsto \eta_\varphi = D \log D\varphi$$

is a bijection, with its inverse given by (5.1.1),

$$\varphi(x) = \frac{\int_0^x e^{\int_0^s \eta} ds}{\int_0^1 e^{\int_0^s \eta} ds}.$$

Using this bijection we identify $\mathcal{D}^3$ with the Banach manifold $([0,1])^2 \times (\mathcal{C}^1([0,1]))^2$ where the identification $\mathcal{D}^3 \longrightarrow ([0,1])^2 \times (\mathcal{C}^1([0,1]))^2$ is given by

$$f = (c, v, f_+, f_-) \longmapsto (c, v, \eta_+, \eta_-) = (c, v, D\log D\varphi_+, D\log D\varphi_-)$$

and φ_+, φ_- are the normalized maps of the two branches f_+, f_-.

Lemma 10.1.1 *The renormalization operator R is not differentiable along the one-parameter family*

$$\mathbb{R} \ni t \mapsto f_t = (c, v, \eta_+, \eta_- + t) \in \mathcal{D}^3_-$$

where $c, v \in [0,1)$, $v > c$, $\eta_+ \in \mathcal{C}^1([0,1])$ and $D\eta_+$ is continuous and nowhere differentiable.

Proof Let φ_+, φ_{t-} be the corresponding diffeomorphisms associated to f_t. Denote by

$$Rf_t := (\tilde{c}(t), \tilde{v}(t), \tilde{\eta}_+(t), \tilde{\eta}_-(t))$$

and we compute $\tilde{\eta}_-(t)$. By the definition of renormalization, the normalized left branch of Rf_t is

$$\tilde{\varphi}_-(t) = A^{-1}_{\left[v\varphi_+(\frac{v-c}{1-c}), v\right]} \circ (f_t)_+ \circ (f_t)_- \circ A_{[0,c]}.$$

Moreover,

$$(f_t)_+ = A_{[0,v]} \circ \varphi_+ \circ A^{-1}_{[c,1]}, \quad (f_t)_- = A_{[v,1]} \circ \varphi_{t-} \circ A^{-1}_{[0,c]}.$$

Notice that by Lemma 3.2.3, for any affine map A,

$$\eta_{A\circ\varphi}(x) = 0 \cdot D\varphi(x) + \eta_\varphi(x) = \eta_\varphi(x), \quad \forall \varphi.$$

Hence

$$\begin{aligned}\tilde{\eta}_-(t)(x) &= \eta_{\tilde{\varphi}_-(t)}(x) = \eta_{\varphi_+ \circ A^{-1}_{[c,1]} \circ A_{[v,1]} \circ \varphi_{t-}}(x) \\ &= \eta_+\left(\frac{(1-v)\varphi_{t-}(x) + v - c}{1-c}\right) \cdot \frac{1-v}{1-c} \cdot D\varphi_{t-}(x) + \eta_-(x) + t.\end{aligned}$$

Since $D\eta_+$ is only continuous and nowhere differentiable, we have $D\eta_+ \notin \mathcal{C}^1([\frac{v-c}{1-c}, 1])$, hence

$$\begin{aligned}\frac{d}{dt}\tilde{\eta}_-(t)(x) = D\eta_+ &\left(\frac{(1-v)\varphi_{t-}(x)+v-c}{1-c}\right)\cdot\frac{1-v}{1-c}\cdot\frac{d}{dt}\varphi_{t-}(x)\\ &\cdot\frac{1-v}{1-c}\cdot D\varphi_{t-}(x)\\ &+\eta_+\left(\frac{(1-v)\varphi_{t-}(x)+v-c}{1-c}\right)\cdot\frac{1-v}{1-c}\cdot\frac{d}{dt}D\varphi_{t-}(x)\\ &+1\notin C^1([0,1]),\end{aligned}$$

where we used that $\frac{d}{dt}\varphi_{t-}$ and $\frac{d}{dt}D\varphi_{t-}$ are continuously differentiable in x. □

Lemma 10.1.2 *The renormalization operator R is not differentiable along the one-parameter family*

$$(0,1)\ni t\mapsto f_t=(c,c+t(1-c),\eta_+,\eta_-)\in\mathcal{D}^3_-$$

where $c\in(0,1)$ and $\eta_+\in C^1([0,1])\backslash C^2([0,1])$.

Proof Let φ_+,φ_- be the corresponding diffeomorphisms associated to f_t. Same as before we denote by $Rf_t:=(\tilde{c}(t),\tilde{v}(t),\tilde{\eta}_+(t),\tilde{\eta}_-(t))$ and we compute $\tilde{\eta}_+(t)$. By the definition of renormalization, the normalized right branch of Rf_t is

$$\tilde{\varphi}_+(t)(x)=A^{-1}_{[0,\varphi_+(t)]}\circ\varphi_+\circ A_{[0,t]}(x)=\frac{\varphi_+(tx)}{\varphi_+(t)}.$$

Hence,

$$\tilde{\eta}_+(t)(x)=\eta_{\tilde{\varphi}_+(t)}(x)=\frac{D^2\tilde{\varphi}_+(t)(x)}{D\tilde{\varphi}_+(t)(x)}=\frac{\frac{t^2}{\varphi_+(t)}D^2\varphi_+(tx)}{\frac{t}{\varphi_+(t)}D\varphi_+(tx)}=t\eta_+(tx),$$

and for $\eta_+\notin C^2([0,1])$,

$$\frac{d}{dt}\tilde{\eta}_+(t)(x)=\eta_+(tx)+txD\eta_+(tx)\notin C^1([0,1]).$$

□

Remark 10.1.3 Observe that the renormalization operator R is also not differentiable in the variable η_- along the family in Lemma 10.1.2. It can be proved by similar computation as in Lemma 10.1.1. Both restriction and composition are responsible for the non-differentiability of R in this case.

10.2 Topological Classes

Given a map $f = (c, c, \varphi_-, \varphi_+) \in \mathcal{D}_0^3$, we define a family

$$\mathcal{F}^f : [0, 1) \ni s \mapsto f_s := (c, s, \varphi_-, \varphi_+) \in \mathcal{D}^3.$$

In particular, $f = f_c$. Note that $\mathcal{F}^f$ is a monotone family in the sense that, if $s_1 < s_2$ then $f_{s_1} \prec f_{s_2}$. By Theorem 2.4.1, the family $\mathcal{F}^f$ is a full family. Moreover, for each irrational number $\rho_0 \in [0, 1)\backslash\mathbb{Q}$, by Remark 2.4.8 there exists a unique $s_0 \in [0, 1)$ dependent on f and ρ_0 such that $\rho(f_{s_0}) = \rho_0$. Hence we have a well-defined function

$$t : \mathcal{D}_0^3 \times ([0, 1)\backslash\mathbb{Q}) \longrightarrow [0, 1), \quad (f, \rho_0) \longmapsto s_0. \tag{10.2.1}$$

Observe that every map $(c, v, \varphi_-, \varphi_+)$ in $\mathcal{D}^3$ belongs to exactly one such full family $\mathcal{F}^f$, where $f = (c, c, \varphi_-, \varphi_+) \in \mathcal{D}_0^3$. Hence, every map in $\mathcal{D}^3$ can be written in a unique way in the form f_v for its corresponding $f \in \mathcal{D}_0^3$ and parameter $v \in [0, 1)$. For every $\rho_0 \in [0, 1)\backslash\mathbb{Q}$, let $\mathcal{W}_{\rho_0}^s$ denote the topological class of rotation number ρ_0 in $\mathcal{D}^3$. More precisely,

$$\mathcal{W}_{\rho_0}^s = \Big\{ f_v \mid f \in \mathcal{D}_0^3,\ v = t(f, \rho_0) \Big\} \subset \mathcal{D}^3.$$

Lemma 10.2.1 *For every $\rho_0 \in [0, 1)\backslash\mathbb{Q}$, $\mathcal{W}_{\rho_0}^s$ is the graph of a C^0 function. Namely, the map*

$$\mathcal{D}_0^3 \ni f \longmapsto t(f, \rho_0) \in [0, 1)$$

is continuous for all irrational number ρ_0.

Proof Fix $f \in \mathcal{D}_0^3$ and let $s_0 = t(f, \rho_0)$. For any $\varepsilon > 0$, let $s_+ = s_0 + \varepsilon$, $s_- = s_0 - \varepsilon$ and $\rho_+ = \rho(f_{s_+})$, $\rho_- = \rho(f_{s_-})$. Since $\rho_0 \notin \mathbb{Q}$, by Proposition 2.4.7 we have $\rho_- < \rho_0 < \rho_+$. Let $\xi = \min\{\,|\rho_0 - \rho_-|,\ |\rho_0 - \rho_+|\,\} > 0$. Recall that the map $f \mapsto \rho(f)$ is continuous in the C^0 topology, see Proposition 2.4.3. Hence, there exists a constant $\delta_- > 0$ such that, for all circle homeomorphisms g with $d(g, f_{s_-}) < \delta_-$ we have $|\rho(g) - \rho(f_{s_-})| < \frac{1}{2}\xi$. By Remark 4.1.1, there exists δ'_- such that, for any $\tilde{f} \in \mathcal{D}_0^3$ with $d_3(f, \tilde{f}) < \delta'_-$,

$$d(f_s, \tilde{f}_s) < \delta_-, \quad \forall s \in [0, 1),$$

and thus

$$|\rho(\tilde{f}_{s_-}) - \rho(f_{s_-})| = |\rho(\tilde{f}_{s_-}) - \rho_-| < \frac{1}{2}\xi.$$

By our choice of ξ,

$$\rho(\tilde{f}_{s_-}) < \rho_- + \frac{1}{2}\xi \leq \rho_0 - \frac{1}{2}\xi < \rho_0.$$

Similarly there exists a constant $\delta'_+ > 0$ such that for all $\tilde{f} \in \mathcal{D}_0^3$ with $d_3(f, \tilde{f}) < \delta'_+$ we have

$$\rho(\tilde{f}_{s_+}) > \rho_+ - \frac{1}{2}\xi \geq \rho_0 + \frac{1}{2}\xi > \rho_0.$$

Let $\delta' = \min\{\delta'_-, \delta'_+\}$. Since $\rho(\tilde{f}_s)$ is monotone increasing in s, see Remark 2.4.6, the above discussion yields that $s_- < t(\tilde{f}, \rho_0) < s_+$ for all $\tilde{f} \in \mathcal{D}_0^3$ with $d_3(f, \tilde{f}) < \delta'$. Hence for all $\tilde{f} \in \mathcal{D}_0^3$ with $d_3(f, \tilde{f}) < \delta'$, we have $|t(\tilde{f}, \rho_0) - s_0| < \varepsilon$. This concludes the proof that $f \mapsto t(f, \rho_0)$ is continuous. □

Definition 10.2.2 We call a topological space L a lamination if it is homeomorphic to a product $M \times \Sigma$, where M is a Banach manifold, and Σ is totally disconnected. A connected component of L is called a leaf of the lamination.

Proposition 10.2.3 *The union of the topological classes,*

$$L = \bigcup_{\rho \in [0,1)\backslash\mathbb{Q}} \mathcal{W}_\rho^s$$

is a lamination. In particular L is homeomorphic to the product $\mathcal{D}_0^3 \times ([0, 1)\backslash\mathbb{Q})$.

Proof We prove that the map

$$\mathcal{D}_0^3 \times ([0, 1)\backslash\mathbb{Q}) \ni (f, \rho_0) \longmapsto f_{t(f,\rho_0)} \in \bigcup_{\rho \in [0,1)\backslash\mathbb{Q}} \mathcal{W}_\rho^s \tag{10.2.2}$$

is a homeomorphism. The map is indeed well-defined as $t(f, \rho_0)$ is a well-defined function. Conversely, since every map in $\bigcup_{\rho \in [0,1)\backslash\mathbb{Q}} \mathcal{W}_\rho^s$ is of the form f_v in a full family $\mathcal{F}^f$, where $f \in \mathcal{D}_0^3$ and $\rho(f_v) \in [0, 1)\backslash\mathbb{Q}$, the map has an inverse

$$f_v \longmapsto (f, \rho(f_v)) \in \mathcal{D}_0^3 \times ([0, 1)\backslash\mathbb{Q}). \tag{10.2.3}$$

This implies that (10.2.2) is a bijection. We are left to show that both directions are continuous.

Observe that (10.2.2) is continuous in variable f by Lemma 10.2.1. For variable ρ first notice that the map $\rho \mapsto t(f, \rho)$ is in essence the inverse of the map $s \mapsto \rho(f_s)$, which is the "devil's staircase" as described in Remark 2.4.8. It is continuous and strictly increasing at irrational values. Observe that the discontinuities of $\rho \mapsto t(f, \rho)$ are exactly at the plateau of the devil's staircase, which are the rational values. Hence $\rho \mapsto t(f,\rho)$ is continuous on $[0,1)\backslash\mathbb{Q}$, and thus (10.2.2) is continuous.

For (10.2.3), let $f_v = (c, v, \varphi_-, \varphi_+)$ be a map in $\bigcup_{\rho\in[0,1)\backslash\mathbb{Q}} \mathcal{W}^s_\rho \subset \mathcal{D}^3$. Observe that for any other map $\tilde{f}_{v'} = (c', v', \psi_-, \psi_+) \in \bigcup_{\rho\in[0,1)\backslash\mathbb{Q}} \mathcal{W}^s_\rho$,

$$d_3(f_v, \tilde{f}_{v'}) = \max\{ d_3(f, \tilde{f}), |v - v'| \} \geq d_3(f, \tilde{f}),$$

hence $f_v \mapsto f$ is continuous. Since the map $f_v \mapsto \rho(f_v)$ is continuous in the distance d by Proposition 2.4.3 and d is controlled by d_3 by Remark 4.1.1, $f_v \mapsto \rho(f_v)$ is continuous in d_3. This implies that the map (10.2.3) is continuous. □

10.3 Stable Set of Fast Renormalization

Definition 10.3.1 Let $f : X \to X$ be a dynamical system where X is equipped with a distance d, and let $x \in X$. The (global) stable set of x is defined as

$$W^s(x) = \{ y \in X \mid d(f^n(x), f^n(y)) \to 0 \text{ as } n \to \infty \}.$$

The following proposition tells us that for the dynamical system of fast renormalization, the stable sets are exactly the topological classes $\mathcal{W}^s_{\rho_0}$ for rotation numbers $\rho_0 \in E_{\lambda_0}$ (see Theorem 8.1.4). In particular, by Lemma 10.2.1, the stable sets $\mathcal{W}^s_{\rho_0}$ for $\rho_0 \in E_{\lambda_0}$ are stable manifolds.

Proposition 10.3.2 *For all $\rho_0 \in E_{\lambda_0}$, the set $\mathcal{W}^s_{\rho_0}$ is the (global) stable set of r_{ρ_0} under fast renormalization. In particular,*

$$d_1(fastR^n f_v, fastR^n r_{\rho_0}) \to 0 \text{ as } n \to \infty, \quad \forall f_v \in \mathcal{W}^s_{\rho_0},$$

and the convergence is exponential.

Proof For any $f_v \in \mathcal{W}^s_{\rho_0} \subset \mathcal{D}^3$, since $\rho(f_v) = \rho_0 \notin \mathbb{Q}$ we have $f_v \in \mathcal{I}^3$. Then by Theorem 7.1.1,

$$d_1(fastR^n f_v, r_{v_{\alpha_n}}) \to 0 \text{ as } n \to \infty.$$

Let $v^*_{\alpha_n}$ denote $fastR^n r_{\rho_0}(0)$, notice that $fastR^n r_{\rho_0} = r_{v^*_{\alpha_n}}$, and thus it suffices to show

$$|v_{\alpha_n} - v^*_{\alpha_n}| \to 0 \text{ exponentially as } n \to \infty.$$

Let us consider the case $fastR^n f_v \in \mathcal{D}^3_-$. The case $fastR^n f_v \in \mathcal{D}^3_+$ can be treated similarly. We use the same notation as in Chap. 8. Observe that

$$v^*_{\alpha_n} = \frac{|L^*_{\alpha_n}| + |R^*_{\alpha_n}| - |r^{l_{\alpha_n}}_{\rho_0}(L^*_{\alpha_n})|}{|L^*_{\alpha_n}| + |R^*_{\alpha_n}|} = \frac{|L^*_{\alpha_n}| + |R^*_{\alpha_n}| - |L^*_{\alpha_n}|}{|L^*_{\alpha_n}| + |R^*_{\alpha_n}|} = \frac{1}{\theta^*_n + 1} \tag{10.3.1}$$

and

$$v_{\alpha_n} = \frac{|L_{\alpha_n}| + |R_{\alpha_n}| - |f^{l_{\alpha_n}}_v(L_{\alpha_n})|}{|L_{\alpha_n}| + |R_{\alpha_n}|}.$$

By Theorem 7.1.1 there exist constants $C > 0$ and $0 < \zeta < 1$ such that

$$(1 - C\zeta^n)|L_{\alpha_n}| \le |f^{l_{\alpha_n}}_v(L_{\alpha_n})| \le (1 + C\zeta^n)|L_{\alpha_n}|.$$

Hence

$$\frac{1 - \theta_n C\zeta^n}{\theta_n + 1} \le v_{\alpha_n} \le \frac{1 + \theta_n C\zeta^n}{\theta_n + 1}. \tag{10.3.2}$$

By (10.3.1) and (10.3.2) we have

$$\begin{aligned} |v_{\alpha_n} - v^*_{\alpha_n}| &\le \left| \frac{1}{\theta_n + 1} - \frac{1}{\theta^*_n + 1} \right| + \left| \frac{\theta_n C\zeta^n}{\theta_n + 1} \right| \\ &= \frac{\theta^*_n}{(\theta_n + 1)(\theta^*_n + 1)} |\Delta_n| + \frac{C\theta_n}{\theta_n + 1}\zeta^n \le |\Delta_n| + C\zeta^n. \end{aligned}$$

Since $\rho(f_v) = \rho_0 \in E_{\lambda_0}$, recall that there exists $\tilde{\lambda} < \lambda_0$ such that

$$\epsilon_n(f_v) \le C \left(\frac{\tilde{\lambda}}{\lambda_0} \right)^n,$$

see the proof of Lemma 8.1.5. Using Proposition 8.2.3 we have that,

$$\begin{aligned} |\Delta_n| &\le C \sum_{\ell \ge n} \epsilon_\ell(f_v) \cdot \left(\frac{2}{3} \right)^{\ell - n} \\ &\le C \left(\frac{2}{3} \right)^{-n} \sum_{\ell \ge n} \left(\frac{2\tilde{\lambda}}{3\lambda_0} \right)^\ell = C \left(\frac{2}{3} \right)^{-n} \cdot \frac{\left(\frac{2\tilde{\lambda}}{3\lambda_0} \right)^n}{1 - \frac{2\tilde{\lambda}}{3\lambda_0}} \\ &\le C \left(\frac{\tilde{\lambda}}{\lambda_0} \right)^n. \end{aligned}$$

Hence $|v_{\alpha_n} - v^*_{\alpha_n}| \to 0$ exponentially as $n \to \infty$. □

10.4 Unstable Part of Fast Renormalization

For any positive integer a, let $\mathcal{D}^3_{a,-}$ and $\mathcal{D}^3_{a,+}$ denote respectively the subsets of $\mathcal{D}^3_-$ and $\mathcal{D}^3_+$ on which $fastRf = R^a f$. Let $\mathcal{W}^-_a = \mathcal{W} \cap \mathcal{D}^3_{a_-}$ and $\mathcal{W}^+_a = \mathcal{W} \cap \mathcal{D}^3_{a_+}$. Recall that for the set of rotations, in Lemma 4.5.1 we defined a map $R : [0, \frac{1}{2}) \cup (\frac{1}{2}, 1) \to [0, 1)$

$$R(\rho) = \begin{cases} \frac{\rho}{1-\rho} & \rho \in [0, \frac{1}{2}) \\ \frac{2\rho-1}{\rho} & \rho \in (\frac{1}{2}, 1) \end{cases},$$

whose derivative is strictly greater than 1. Hence for rotations the fast renormalization $fastR$ is represented by R^a on intervals $\mathcal{W}^-_a, \mathcal{W}^+_a$, with derivative greater than 1.

Proposition 10.4.1 *Let $\mathcal{W}_0 := \bigcup_{a \geq 1} \mathcal{W}^-_a \cup \mathcal{W}^+_a$. Fast renormalization is expanding on $\mathcal{W}_0$. More precisely,*

$$fastR : \mathcal{W}_0 \to \mathcal{W}$$

satisfies

$$\min_{\rho \in \mathcal{W}_0} \frac{d}{d\rho} fastR(\rho) > 1.$$

10.5 Exercises

Exercise 10.5.1 Show that each branch $fastR : \mathcal{D}^3_{a,\pm} \to \mathcal{D}^3$ of fast renormalization is onto. Compare Fig. 10.1.

References

[Arn61] Arnold, V.I.: Small denominators. I. Mapping the circle onto itself. Izv. Akad. Nauk SSSR Ser. Mat. **25**, 21–86 (1961)

[Arn63] Arnold, V.I.: Proof of a theorem of A. N. Kolmogorov on the preservation of conditionally periodic motions under a small perturbation of the Hamiltonian. Uspehi Mat. Nauk **18**(5)(113), 13–40 (1963)

[Arn83] Arnold, V.I.: In: M. Levi (ed.) Geometrical Methods in the Theory of Ordinary Differential Equations. Grundlehren der Mathematischen Wissenschaften, vol. 250, p. xi+334. Springer, New York (1983). Translated from the Russian by Joseph Szücs

[Avi13] Avila, A.: On rigidity of critical circle maps. Bull. Braz. Math. Soc. (N.S.) **44**(4), 611–619 (2013)

[AL11] Avila, A., Lyubich, M.: The full renormalization horseshoe for unimodal maps of higher degree: exponential contraction along hybrid classes. Publ. Math. Inst. Hautes Études Sci. **114**, 171–223 (2011)

[ALM03] Avila, A., Lyubich, M., and W. de Melo: Regular or stochastic dynamics in real analytic families of unimodal maps. Invent. Math. **154**(3), 451–550 (2003)

[AM03] Avila, A., Moreira, C.G.: Statistical properties of unimodal maps: smooth families with negative Schwarzian derivative. Astérisque **286** (2003). Geometric methods in dynamics. I, xviii, 81–118.

[AM05a] Avila, A., Moreira, C.G.: Phase-parameter relation and sharp statistical properties for general families of unimodal maps. In: Geometry and Dynamics, vol. 389, pp. 1–42. Contemp. Math. Amer. Math. Soc., Providence (2005)

[AM05b] Avila, A., Moreira, C.G.: Statistical properties of unimodal maps: physical measures, periodic orbits and pathological laminations. Publ. Math. Inst. Hautes Études Sci. **101**, 1–67 (2005)

[AM05c] Avila, A., Moreira, C.G.: Statistical properties of unimodal maps: the quadratic family. Ann. Math. (2) **161**(2), 831–881 (2005)

[BG94] Barrow-Green, J.: Oscar II's prize competition and the error in Poincaré's memoir on the three body problem. Arch. Hist. Exact Sci. **48**(2), 107–131 (1994)

[BC91] Benedicks, M., Carleson, L.: The dynamics of the Hénon map. Ann. Math. (2) **133**(1), 73–169 (1991)

[BP23] Benedicks, M., Palmisano, L.: Coexistence phenomena in the Hénon family. Bull. Braz. Math. Soc. (N.S.) **54**(3), Paper No. 42, 42 (2023)

[Ber16] Berger, P.: Generic family with robustly infinitely many sinks. Invent. Math. **205**(1), 121–172 (2016)

Y. Dong et al., *One-Dimensional Dynamics*, Lecture Notes of the Unione Matematica Italiana 30, https://doi.org/10.1007/978-3-031-94882-4

[BMT03] Birkhoff, G., Martens, M., Tresser, C.: On the scaling structure for period doubling. Astérisque **286** (2003). Geometric methods in dynamics. I, xix, 167–186.

[BL89] Blokh, A.M., Lyubich, M.Y.: Nonexistence of wandering intervals and structure of topological attractors of one-dimensional dynamical systems. II. The smooth case. Ergodic Theory Dynam. Syst. **9**(4), 751–758 (1989)

[BS02] Brin, M., Stuck, G.: Introduction to Dynamical Systems, xii+240. Cambridge University Press, Cambridge (2002)

[Cha+09] Chandramouli, V.V.M.S., Martens, M., de Melo, W., Tresser, C.P.: Chaotic period doubling. Ergodic Theory Dynam. Syst. **29**(2), 381–418 (2009)

[Cla14] Clark, T.: Regular or stochastic dynamics in families of higher-degree unimodal maps. Ergodic Theory Dynam. Syst. **34**(5), 1538–1566 (2014)

[CS23] Clark, T., van Strien, S.: Conjugacy classes of real analytic one-dimensional maps are analytic connected manifolds (2023, Apr. 3). arXiv: `2304.00883v1 [math.DS]`

[CEK81a] Collet, P., Eckmann, J.-P., Koch, H.: On universality for area-preserving maps of the plane. Phys. D **3**(3), 457–467 (1981)

[CEK81b] Collet, P., Eckmann, J.-P., Koch, H.: Period doubling bifurcations for families of maps on $\mathbf{R}^n$. J. Stat. Phys. **25**(1), 1–14 (1981)

[Cop52] Copernicus, N.: On the Revolutions of the Heavenly Spheres. Great Books of the Western World, vol. 16, pp. 497–838. Encyclopaedia Britannica, Chicago (1952)

[CPT20] Crovisier, S., Pujals, E., Tresser, C.: Mildly dissipative diffeomorphisms of the disk with zero entropy (2020, May 28). arXiv: `2005.14278v3 [math.DS]`

[Dav91] Davis, G.J.: Infinitely many co-existing sinks from degenerate homoclinic tangencies. Trans. Amer. Math. Soc. **323**(2), 727–748 (1991)

[DCLM05] De Carvalho, A., Lyubich, M., Martens, M.: Renormalization in the Hénon family. I. Universality but non-rigidity. J. Stat. Phys. **121**(5–6), 611–669 (2005)

[Den32] Denjoy, A.: Sur les courbes définies par les équations différentielles à la surface du tore. J. Math. Pures Appl. **9**(11), 333–375 (1932)

[Dov+06] Dovgoshey, O., Martio, O., Ryazanov, V., Vuorinen, M.: The Cantor function. Expos. Math. **24**(1), 1–37 (2006)

[EKW82] Eckmann, J.-P., Koch, H., Wittwer, P.: Existence of a fixed point of the doubling transformation for area-preserving maps of the plane. Phys. Rev. A (3) **26**(1), 720–722 (1982)

[EKW84] Eckmann, J.-P., Koch, H., Wittwer, P.: A computer-assisted proof of universality for area-preserving maps. Mem. Amer. Math. Soc. **47**(289), vi+122 (1984)

[Eps86] Epstein, H.: New proofs of the existence of the Feigenbaum functions. Comm. Math. Phys. **106**(3), 395–426 (1986)

[Far99] de Faria, E.: Asymptotic rigidity of scaling ratios for critical circle mappings. Ergodic Theory Dynam. Syst. **19**(4), 995–1035 (1999)

[FM99] de Faria, E., de Melo, W.: Rigidity of critical circle mappings. I. J. Eur. Math. Soc. (JEMS) **1**(4), 339–392 (1999)

[FM00] de Faria, E., de Melo, W.: Rigidity of critical circle mappings. II. J. Amer. Math. Soc. **13**(2), 343–370 (2000)

[FMP06] de Faria, E., de Melo, W., Pinto, A.: Global hyperbolicity of renormalization for C^r unimodal mappings. Ann. Math. (2) **164**(3), 731–824 (2006)

[Fei78] Feigenbaum, M.J.: Quantitative universality for a class of nonlinear transformations. J. Stat. Phys. **19**(1), 25–52 (1978)

[Fei79] Feigenbaum, M.J.: The universal metric properties of nonlinear transformations. J. Stat. Phys. **21**(6), 669–706 (1979)

[FKS82] Feigenbaum, M.J., Kadanoff, L.P., Shenker, S.J.: Quasiperiodicity in dissipative systems: a renormalization group analysis. Phys. D **5**(2–3), 370–386 (1982)

[GJ09] Gaidashev, D., Johnson, T.: Dynamics of the universal area-preserving map associated with period doubling: hyperbolic sets. Nonlinearity **22**(10), 2487–2520 (2009)

[GJ16] Gaidashev, D., Johnson, T.: Spectral properties of renormalization for area-preserving maps. Discrete Contin. Dyn. Syst. **36**(7), 3651–3675 (2016)

[GJM16] Gaidashev, D., Johnson, T., Martens, M.: Rigidity for infinitely renormalizable area-preserving maps. Duke Math. J. **165**(1), 129–159 (2016)

[Gal89] Galilei, G.: Sidereus nuncius, or, The Sidereal Messenger. Translated from the Latin and with an introduction, conclusion and notes by Albert Van Helden, pp. xii+127. University of Chicago Press, Chicago (1989)

[GST89] Gambaudo, J.-M., van Strien, S., Tresser, C.: Hénon-like maps with strange attractors: there exist C^∞ Kupka-Smale diffeomorphisms on S^2 with neither sinks nor sources. Nonlinearity **2**(2), 287–304 (1989)

[GU21] Ghazouani, S., Ulcigrai, C.: A priori bounds for GIETs, affine shadows and rigidity of foliations in genus 2 (2021, June 7). arXiv: `2106.03529v1 [math.DS]`

[GST08] Gonchenko, S.V., Shilnikov, L.P., and Turaev, D.V.: On dynamical prop- erties of multidimensional diffeomorphisms from Newhouse regions. I. Nonlinearity **21**(5), 923–972 (2008)

[GM17] Guarino, P., de Melo, W.: Rigidity of smooth critical circle maps. J. Eur. Math. Soc. (JEMS) **19**(6), 1729–1783 (2017)

[GMM18] Guarino, P., Martens, M., de Melo, W.: Rigidity of critical circle maps. Duke Math. J. **167**(11), 2125–2188 (2018)

[Guc79] Guckenheimer, J.: Sensitive dependence to initial conditions for one-dimensional maps. Comm. Math. Phys. **70**(2), 133–160 (1979)

[Guc87] Guckenheimer, J.: Limit sets of S-unimodal maps with zero entropy. Comm. Math. Phys. **110**(4), 655–659 (1987)

[GH83] Guckenheimer, J., Holmes, P.: Nonlinear Oscillations, Dynamical Systems, and Bifurcations of Vector Fields. Applied Mathematical Sciences, vol. 42, pp. xvi+453. Springer, New York (1983)

[Hal81] Hall, G.R.: A C^∞ Denjoy counterexample. Ergodic Theory Dynam. Syst. **1**(3), 261–272 (1981)

[HW08] Hardy, G.H., Wright, E.M.: An Introduction to the Theory of Numbers. Sixth. Revised by D. R. Heath-Brown and J. H. Silverman, With a foreword by Andrew Wiles, pp. xxii+621. Oxford University Press, Oxford (2008)

[HLM12] Hazard, P.E., Lyubich, M., Martens, M.: Renormalizable Hénon-like maps and unbounded geometry. Nonlinearity **25**(2), 397–420 (2012)

[Her79] Herman, M.-R.: Sur la conjugaison différentiable des difféomorphismes du cercle à des rotations. Inst. Hautes Études Sci. Publ. Math. **49**, 5–233 (1979)

[KH95] Katok, A., Hasselblatt, B.: Introduction to the Modern Theory of Dynamical Systems. Encyclopedia of Mathematics and its Applications, vol. 54, pp. xviii+802. Cambridge University Press, Cambridge (1995). With a supplementary chapter by Katok and Leonardo Mendoza

[KO89] Katznelson, Y., Ornstein, D.: The differentiability of the conjugation of certain diffeomorphisms of the circle. Ergodic Theory Dynam. Syst. **9**(4), 643–680 (1989)

[KS87] Khanin, K.M., Sinaĭ, Y.G.: A new proof of M. Herman's theorem. Comm. Math. Phys. **112**(1), 89–101 (1987)

[KT07] Khanin, K., Teplinsky, A.: Robust rigidity for circle diffeomorphisms with singularities. Invent. Math. **169**(1), 193–218 (2007)

[KY06] Khmelev, D., Yampolsky, M.: The rigidity problem for analytic critical circle maps. Mosc. Math. J. **6**(2), 317–351, 407 (2006)

[Kol54] Kolmogorov, A.N.: On conservation of conditionally periodic motions for a small change in Hamilton's function. Dokl. Akad. Nauk SSSR (N. S.) **98**, 527–530 (1954)

[Lan82] Lanford III, O.E.: A computer-assisted proof of the Feigenbaum conjectures. Bull. Amer. Math. Soc. (N. S.) **6**(3), 427–434 (1982)

[Lan87] Lanford III, O.E.: Renormalization group methods for critical circle mappings with general rotation number. In: VIIIth International Congress on Mathematical Physics (Marseille, 1986), pp. 532–536. World Scientific Publishing, Singapore (1987)

[Lan88] Lanford III, O.E.: Renormalization group methods for circle mappings. In: Nonlinear Evolution and Chaotic Phenomena (Noto, 1987). NATO Adv. Sci. Inst. Ser. B: Phys., vol. 176, pp. 25–36. Plenum, New York (1988)

[Lla01] de la Llave, R.: A tutorial on KAM theory. Smooth ergodic theory and its applications (Seattle, WA, 1999). Proc. Sympos. Pure Math., vol. 69, pp. 175–292. American Mathematical Society, Providence (2001)

[LM11a] Lyubich, M., Martens, M.: Renormalization in the Hénon family, II: the heteroclinic web. Invent. Math. **186**(1), 115–189 (2011)

[LM11b] Lyubich, M., Martens, M.: Renormalization of Hénon maps. In: Dynamics, Games and Science. I. Springer Proc. Math., vol. 1, pp. 597–618. Springer, Heidelberg (2011)

[LM21] Lyubich, M., Martens, M.: Probabilistic universality in two-dimensional dynamics. Comm. Math. Phys. **383**(3), 1295–1359 (2021)

[Lyu89] Lyubich, M.Y.: Nonexistence of wandering intervals and structure of topological attractors of one-dimensional dynamical systems. I. The case of negative Schwarzian derivative. Ergodic Theory Dynam. Syst. **9**(4), 737–749 (1989)

[Lyu00] Lyubich, M.: Dynamics of quadratic polynomials. III. Parapuzzle and SBR measures. Astérisque **261** (2000). Géométrie complexe et systèmes dynamiques (Orsay, 1995), xii–xiii, 173–200

[Lyu02] Lyubich, M.: Almost every real quadratic map is either regular or stochastic. Ann. Math. (2) **156**(1), 1–78 (2002)

[LM93] Lyubich, M., Milnor, J.: The Fibonacci unimodal map. J. Amer. Math. Soc. **6**(2), 425–457 (1993)

[Mn87] Manñé, R.: Ergodic Theory and Differentiable Dynamics. Ergebnisse der Mathematik und ihrer Grenzgebiete (3) [Results in Mathematics and Related Areas (3)], vol. 8, pp. xii+317. Springer, Berlin (1987). Translated from the Portuguese by Silvio Levy

[MP19] Marchese, L., Palmisano, L.: Full families of generalized interval exchange transformations. Nonlinearity **32**(1), 110–142 (2019)

[Mar98] Martens, M.: The periodic points of renormalization. Ann. Math. (2) **147**(3), 543–584 (1998)

[MM01] Martens, M., de Melo, W.: Universal models for Lorenz maps. Ergodic Theory Dynam. Syst. **21**(3), 833–860 (2001)

[MP22] Martens, M., Palmisano, L.: Invariant manifolds for non-differentiable operators. Trans. Amer. Math. Soc. **375**(2), 1101–1169 (2022)

[MW14] Martens, M., Winckler, B.: On the hyperbolicity of Lorenz renormalization. Comm. Math. Phys. **325**(1), 185–257 (2014)

[MW16] Martens, M., Winckler, B.: Instability of renormalization (2016, Sept. 14). arXiv: `1609.04473v3 [math.DS]`

[Mar+90] Martens, M., van Strien, S., de Melo, W., Mendes, P.: On Cherry flows. Ergodic Theory Dynam. Syst. **10**(3), 531–554 (1990)

[MMS92] Martens, M., de Melo, W., van Strien, S.: Julia-Fatou-Sullivan theory for real one-dimensional dynamics. Acta Math. **168**(3–4), 273–318 (1992)

[McM94] McMullen, C.T.: Complex Dynamics and Renormalization. Annals of Mathematics Studies, vol. 135, pp. x+214. Princeton University Press, Princeton (1994)

[McM98] McMullen, C.T.: Rigidity and inflexibility in conformal dynamics. In: Proceedings of the International Congress of Mathematicians, Vol. II (Berlin, 1998), pp. 841–855 (1998)

[Mel87] de Melo, W.: A finiteness problem for one-dimensional maps. Proc. Amer. Math. Soc. **101**(4), 721–727 (1987)

[MP99] de Melo, W., Pinto, A.A.: Rigidity of C^2 infinitely renormalizable unimodal maps. Comm. Math. Phys. **208**(1), 91–105 (1999)

[MS88] de Melo, W., van Strien, S.: One-dimensional dynamics: the Schwarzian derivative and beyond. Bull. Amer. Math. Soc. (N.S.) **18**(2), 159–162 (1988)

[MS89] de Melo, W., van Strien, S.: A structure theorem in one-dimensional dynamics. Ann. Math. (2) **129**(3), 519–546 (1989)

[MS93] de Melo, W., van Strien, S.: One-dimensional dynamics. In: Ergebnisse der Mathematik und ihrer Grenzgebiete (3) [Results in Mathematics and Related Areas (3)], vol. 25, xiv+605. Springer, Berlin (1993)

[MT88] Milnor, J., Thurston, W.: On iterated maps of the interval. In: Dynamical Systems (College Park, MD, 1986–87). Lecture Notes in Math., vol. 1342, pp. 465–563. Springer, Berlin (1988)

[Mos62] Moser, J.: On invariant curves of area-preserving mappings of an annulus. Nachr. Akad. Wiss. Göttingen Math.-Phys. Kl. II **1962**, 1–20 (1962)

[Mos66] Moser, J.: A rapidly convergent iteration method and non-linear differential equations. II. Ann. Scuola Norm. Sup. Pisa Cl. Sci. (3) **20**, 499–535 (1966)

[New74] Newhouse, S.E.: Diffeomorphisms with infinitely many sinks. Topology **13**, 9–18 (1974)

[New99] Newton, I.: The Principia: Mathematical Principles of Natural Philosophy. A new translation by I. Bernard Cohen and Anne Whitman, assisted by Julia Budenz, Preceded by "A guide to Newton's Principia" by Cohen, pp. xviii+974. University of California Press, Berkeley (1999)

[Öst+83] Östlund, S., Rand, D., Sethna, J., Siggia, E.: Universal properties of the transition from quasiperiodicity to chaos in dissipative systems. Phys. D **8**(3), 303–342 (1983)

[PM82] Palis Jr., J., de Melo, W.: Geometric Theory of Dynamical Systems. An introduction, Translated from the Portuguese by A. K. Manning, pp. xii+198. Springer, New York (1982)

[PT93] Palis, J., Takens, F.: Hyperbolicity and Sensitive Chaotic Dynamics at Homoclinic Bifurcations. Fractal Dimensions and Infinitely Many Attractors. Cambridge Studies in Advanced Mathematics, vol. 35, x+234. Cambridge University Press, Cambridge (1993)

[Pal15] Palmisano, L.: A Denjoy counterexample for circle maps with an half-critical point. Math. Z. **280**(3–4), 749–758 (2015)

[Pal19] Palmisano, L.: Cherry flows with non-trivial attractors. Fundam. Math. **244**(3), 243–253 (2019)

[Pet83] Petersen, K.: Ergodic Theory. Cambridge Studies in Advanced Mathematics, vol. 2, xii+329. Cambridge University Press, Cambridge (1983)

[Poi90] Poincaré, H.: Sur le probléme des trois corps et les équations de la dynamique. Acta Math. **13**, 1–270 (1890)

[Poi92] Poincaré, H.: In: Gauthier Villars, P. (ed.) Les Méthodes Nouvelles de la Mécanique Céleste IIII. Gauthier Villars, Paris (1892–1899)

[Poi93] Poincaré, H.: Mémoire sur les courbes définies par une équation différentielle. Reprints of the originals from 1856 through 1921, p. 540. Éditions Jacques Gabay, Sceaux (1993)

[Pös01] Pöschel, J.: A lecture on the classical KAM theorem. In: Smooth Ergodic Theory and Its Applications (Seattle, WA, 1999). Proc. Sympos. Pure Math., vol. 69, pp. 707–732. American Mathematical Society, Providence (2001)

[Ran88a] Rand, D.A.: Global phase space universality, smooth conjugacies and renormalisation. I. The $C^{1+\alpha}$ case. Nonlinearity **1**(1), 181–202 (1988)

[Ran88b] Rand, D.: Universality and renormalisation in dynamical systems. In: New Directions in Dynamical Systems. London Math. Soc. Lecture Note Ser., vol. 127, pp. 1–56. Cambridge University Press, Cambridge (1988)

[Rob83] Robinson, C.: Bifurcation to infinitely many sinks. Comm. Math. Phys. **90**(3), 433–459 (1983)

[She82] Shenker, S.J.: Scaling behavior in a map of a circle onto itself: empirical results. Phys. D **5**(2–3), 405–411 (1982)

[Shu87] Shub, M.: Global Stability of Dynamical Systems. With the collaboration of Albert Fathi and Rémi Langevin, Translated from the French by Joseph Christy, pp. xii+150. Springer, New York (1987)

[Sma65] Smale, S.: Diffeomorphisms with many periodic points. In: Differential and Combinatorial Topology (A Symposium in Honor of Marston Morse), pp. 63–80. Princeton University Press, Princeton (1965)

[Sma80] Smale, S.: The mathematics of time. In: Essays on Dynamical Systems, Economic Processes, and Related Topics, vi+151. Springer, New York (1980)

[Sta88] Stark, J.: Smooth conjugacy and renormalisation for diffeomorphisms of the circle. Nonlinearity **1**(4), 541–575 (1988)

[Str81] van Strien, S.J.: On the bifurcations creating horseshoes. In: Dynamical Systems and Turbulence, Warwick 1980 (Coventry, 1979/1980). Lecture Notes in Math., vol. 898, pp. 316–351. Springer, Berlin (1981)

[Sul92] Sullivan, D.: Bounds, Quadratic Differentials, and Renormalization Conjectures. American Mathematical Society Centennial Publications (Providence, RI, 1988), vol. II, pp. 417–466. American Mathematical Society, Providence (1992)

[TC78] Tresser, C., Coullet, P.: Itérations d'endomorphismes et groupe de renormalisation. C. R. Acad. Sci. Paris Sér. A-B **287**(7), A577–A580 (1978)

[Tro07] Troy, S.: The Double Pendulum. University of Washington Department of Physics (2007). https://www.youtube.com/watch?v=U39RMUzCjiU

[VSK84] Vul, E.B., Sinaĭ, Y.G., Khanin, K.M.: Feigenbaum universality and thermodynamic formalism. Uspekhi Mat. Nauk. **39**(3)(237), 3–37 (1984)

[Yam99] Yampolsky, M.: Complex bounds for renormalization of critical circle maps. Ergodic Theory Dynam. Syst. **19**(1), 227–257 (1999)

[Yam01] Yampolsky, M.: The attractor of renormalization and rigidity of towers of critical circle maps. Comm. Math. Phys. **218**(3), 537–568 (2001)

[Yam02] Yampolsky, M.: Hyperbolicity of renormalization of critical circle maps. Publ. Math. Inst. Hautes Études Sci. **96**, 1–41 (2002)

[Yam03] Yampolsky, M.: Renormalization horseshoe for critical circle maps. Comm. Math. Phys. **240**(1–2), 75–96 (2003)

[Yoc84a] Yoccoz, J.-C.: Conjugaison différentiable des difféomorphismes du cercle dont le nombre de rotation vérifie une condition diophantienne. Ann. Sci. École Norm. Sup. (4) **17**(3), 333–359 (1984)

[Yoc84b] Yoccoz, J.-C.: Il n'y a pas de contre-exemple de Denjoy analytique. C. R. Acad. Sci. Paris Sér. I Math. **298**(7), 141–144 (1984)

[Yoc95] Yoccoz, J.-C.: Théorème de Siegel, nombres de Bruno et polynômes quadratiques. Astérisque **231**, 3–88 (1995). Petits diviseurs en dimension 1

LECTURE NOTES OF THE UNIONE MATEMATICA ITALIANA

Springer

Editors in Chief: Ciro Ciliberto and Susanna Terracini

Editorial Policy

1. The UMI Lecture Notes aim to report new developments in all areas of mathematics and their applications - quickly, informally and at a high level. Mathematical texts analysing new developments in modelling and numerical simulation are also welcome.
2. Manuscripts should be submitted to
   Redazione Lecture Notes U.M.I.
   umi@dm.unibo.it
   and possibly to one of the editors of the Board informing, in this case, the Redazione about the submission. In general, manuscripts will be sent out to external referees for evaluation. If a decision cannot yet be reached on the basis of the first 2 reports, further referees may be contacted. The author will be informed of this. A final decision to publish cab be made only on the basis of the complete manuscript, however a refereeing process leading to a preliminary decision can be based on a prefinal or incomplete manuscript. The strict minimum amount of material that will be considered should include a detailed outline describing the planned contents of each chapter, a bibliography and several sample chapters.
3. Manuscripts should in general be submitted in English. Final manuscripts should contain at least 100 pages of mathematical text and should always include
   - a table of contents;
   - an informative introduction, with adequate motivation and perhaps some historical remarks: it should be accessible to a reader not intimately familiar with the topic treated;
   - a subject index: as a rule this is genuinely helpful for the reader.
4. For evaluation purposes, please submit manuscripts in electronic form, preferably as pdf- or zipped ps-files. Author are asked, if their manuscript is accepted for publication, to use the LaTeX2e style files available from Springer's web-server at
   ftp://ftp.springer.de/pub/tex/latex/svmonot1/ for monographs
   and at
   ftp://ftp.springer.de/pub/tex/latex/svmultt1/ for multi-authored volumes.
5. Authors receive a total of 50 free copies of their volume, but no royalties. They are entitled to a discount of 33.3 % on the price of Springer books purchased for their personal use, if ordering directly from Springer.
6. Commitment to publish is made by letter of intent rather than by signing a formal contract. Springer-Verlag secures the copyright for each volume. Authors are free to reuse material contained in their LNM volumes in later publications: A brief written (or e-mail) request for formal permission is sufficient.